Routine Lab Values

Made Ridiculously Simple

Stephen Goldberg, M.D.
Professor Emeritus
University of Miami
Miller School of Medicine
Miami FL

MEDMASTER

ISBN #978-1-962445-29-0
eISBN #978-1-962445-30-6

Made in the United States of America

Published by Medmaster, Inc.

Cover by Jordan Collver and Owen Watts

Works by Stephen Goldberg

Books:
Clinical Neuroanatomy Made Ridiculously Simple
The Four-Minute Neurologic Exam
Goldberg's Brain Model
Ophthalmology Made Ridiculously Simple
Jonah: The Anatomy of the Soul
The Jonah Principle: The Basis for Human and Machine Consciousness
Consciousness, Information, and Meaning: The Origin of the Mind
Consciousness: How the Mind Arises from the Brain
Anatomy of the Soul: Mind, God, and the Afterlife
Consciousness Made Ridiculously Simple
Is the Mind Immortal?: The Origin of Consciousness
Clinical Anatomy Made Ridiculously Simple
Clinical Physiology Made Ridiculously Simple
Clinical Biochemistry Made Ridiculously Simple
Clinical Genetics Made Ridiculously Simple
Clinical Biostatistics and Epidemiology Made Ridiculously Simple
LiveCode Lite: Computer Programming Made Ridiculously Simple
Clinical Hematology Made Ridiculously Simple
War Against the Germs: Epidemics, Microorganisms, and Biowarfare
Endocrinology Made Ridiculously Simple
Gastroenterology Made Ridiculously Simple
Neurology Made Ridiculously Simple
Med'Toons

Computer Programs:
Differential Diagnosis
Atlas of Human Diseases
Neurologic Localization
Atlas of Ophthalmology
Atlas of Microbiology
Atlas of Dermatology
Atlas of Orthopedics
Medsearcher

Preface

I wrote this book for medical, nursing, and physician assistant students, primary care physicians, and other health professionals who frequently encounter abnormal values on routine laboratory screening tests—such as the complete blood count (CBC), metabolic panel, coagulation studies, kidney and liver function tests, lipid profile, thyroid function tests, and urinalysis. When an abnormality appears unexpectedly, what does it mean?

For diagnostic (non-screening) tests, the interpretation is usually more straightforward, as these tests are ordered with a specific clinical question in mind. For example, serum troponins are obtained to evaluate for myocardial damage. In contrast, routine screening tests are often ordered without a particular diagnosis in question, making it more challenging to interpret isolated abnormalities, which may arise from a wide range of possible causes.

This book is organized by type of screening test. Each chapter includes tables listing potential causes of elevated or decreased values, along with the underlying pathophysiologic reasoning.

I hope this serves as a useful reference for the day-to-day interpretation of standard screening tests.

I am grateful to Michael A. Chizner, MD, FACP, FACC—Director of Medical Education Emeritus at Broward Health, Clinical Professor of Medicine at the University of Miami Miller School of Medicine, and author of *Clinical Cardiology Made Ridiculously Simple*—for his helpful suggestions. I also thank Phyllis Goldenberg for her excellent proofreading and advice.

Stephen Goldberg, MD

Contents

Introduction

In general, if a measured substance is one that is normally ingested, an elevated value may result from excessive intake or impaired elimination—whether by the kidneys, liver, or other tissues. If the substance is normally produced by the body, the elevation may be due to overproduction or inadequate removal. Alternatively, the total amount may be normal, but the value appears elevated due to concentration effects, such as in dehydration.

Similarly, if a lab value is low, it may be due to insufficient intake (from poor diet or malabsorption), excessive loss, or decreased production. Alternatively, the substance may be present in normal amounts, but the value appears low due to hemodilution.

The term "**abnormal**" can have two meanings. It may indicate a pathological condition, or it may simply reflect a value that falls outside the statistical range of 95% of the range must be interpreted in context, alongside the patient's history, physical examination, and other test results. Furthermore, reference ranges can vary by laboratory, and an "abnormal" value must be interpreted using the range specific to that lab. Normal values may also vary by age, sex, or ethnicity. Sometimes, a single abnormal result may be due to a lab error and should be repeated for confirmation. A summary of generally accepted normal ranges is provided in the Appendix.

It is important to understand what screening laboratory tests can reveal. For example, a **complete blood count (CBC)** can indicate anemia or infection. A lipid panel reflects cardiovascular risk. Liver, kidney, and thyroid function tests help assess organ performance. A high **blood glucose** level may suggest diabetes before symptoms emerge. Lab tests also help monitor therapeutic responses and detect potential risks. For instance, electrolyte disturbances such as low potassium (K^+) or magnesium (Mg^{2+}) may predispose to ventricular arrhythmias, while an elevated thyroxine (T_4) level may be a clue to new-onset heart failure, supraventricular tachycardia, or atrial fibrillation.

Ultimately, lab tests do not function in isolation. They must be interpreted as part of the clinical whole—in the context of the patient's symptoms, physical findings, and other laboratory results. Always consider the big picture.

1

The Complete Blood Count (CBC)

Sometimes, the first indication of what is wrong when the patient is ill is the routine **Complete Blood Count (CBC)** (**Fig. 1-1**).

FIGURE 1-1 NORMAL CBC VALUES IN HEMATOLOGY	
RBC	4.7-6.1 M/mcL
WBC	4.8-10.8 K/mcL
Hgb	14.0-18.0 g/dL
Hct	42-52%
MCV	80-100 fL
RDW	11.5-14.5 %
Iron	60-170 mcg/dL
Ferritin	12-300 ng/mL
TIBC	240-450 mcg/dL
Neutrophils	1.8-7.8 K/mcL (40-60%)
Lymphocytes	1.0-4.8 K/mcL (20-40%)
Monocytes	0-0.8 K/mcL (2-8%)

Eosinophils	0-0.45 K/mcL (1-4%)
Basophils	0-0.2 K/mcL (0.5-1%)
Platelets	150-450 K/mcL
MPV	7.4-11.0 fL
PT	11-13.5 seconds
PTT	25-35 seconds
dL = deciliter; **L**= Liter; **mcg** = micrograms; **mcL** = microliter; **ng** = nanograms	
Hct = hematocrit; **Hgb** = hemoglobin; **MCV** = mean corpuscular volume; **MPV** = mean platelet volume; **PT** = prothrombin time; **PTT** = partial thromboplastin time; **RDW** = Red cell distribution width	
Note: Normal values vary in different labs and with age. Newborns and infants generally have higher hemoglobin, hematocrit, RBC, and WBC counts. In general, RBC counts tend to be higher in men than in women, while platelet counts tend to be higher in women. People of African ancestry tend to display lower Hgb, WBC, and platelet counts than white persons.	

Direct visual examination of the blood smear is also useful. It can detect abnormal cell morphologies (**atypical cells**) that are noted more easily by the human eye than by a lab machine.

Red Blood Cells (RBCs)

For body cells to receive enough oxygen, there needs to be enough functional hemoglobin in the blood. **Anemia** is a condition where there is either a decrease in the number of red blood cells or the red cells do not have enough normal-functioning hemoglobin. In **polycythemia**, there are too many red cells.

RBC Count

Figure 1-2 lists clinical conditions that can change the red blood cell (RBC) count.

FIGURE 1-2 RBC COUNT (RBC)
INCREASED RBC COUNT can be due to: ***Overproduction of RBCs*** • *Hypoxia* (oxygen deficiency), as in moving to a high altitude, cyanotic heart disease, chronic obstructive pulmonary disease, sleep apnea. The stimulus to red cell production is the secretion of *erythropoietin* by the kidney in response to hypoxia.

- *Renal cell carcinoma* that secretes excess erythropoietin
- *Polycythemia vera*, a cancer of the bone marrow, where the marrow produces too many red blood cells

Artifact of dehydration: The total number of RBCs is normal but appears elevated in a blood sample because of their increased concentration (e.g., in diarrhea, burns, excess sweating).

DECREASED RBC COUNT can be due to:
Decreased production of RBCs

- *Poor dietary intake*, or *malabsorption*. The necessary ingredients for making RBCs (e.g., iron, folic acid, vitamin B12) are deficient.
- *Bone marrow compromise* (e.g., leukemia, aplastic anemia)
- *Anemia of chronic disease* (iron can't be utilized to create new RBCs although iron stores are normal or high)

Increased loss of RBCs

- *Hemorrhage*
- *Hemolytic anemia* of numerous causes
- *Burns, infections, frostbite*

Artifact of hemodilution (the total number of red cells may be normal but are diluted with excess body fluid).

Hematocrit (Hct) and Hemoglobin (Hgb)

The **hematocrit** (**Hct**) is a percentage ratio of the *volume of red blood cells* compared with the total volume of blood, as measured in an anticoagulated test tube after centrifugation (about 45% in **Fig. 1-3**).

The **hemoglobin** (**Hgb**) is the *amount of hemoglobin* in the blood as expressed in grams per deciliter (g/dl). In practice, the Hct and Hgb are roughly equivalent in their diagnostic meaning and can be used interchangeably. The conditions that cause an increase or decrease in Hct/Hgb essentially parallel those of increased or decreased RBC count, although it is possible to have an elevated RBC count along with a low Hgb/Hct. For example, in beta-thalassemia trait, the marrow produces a normal or increased number of RBCs. However, the RBCs are small (microcytic), and the amount of hemoglobin in the cells is decreased, as manifest as a low Hgb/Hct.

Mean Corpuscular Hemoglobin (**MCH**) is a supplementary test that give the average **amount** of hemoglobin per red blood cell:

$$MCH = Hgb/RBC\ count \times 10$$

A low MCH suggests that an anemia is microcytic, where the RBCs are smaller than normal (e.g., iron deficiency or thalassemia).

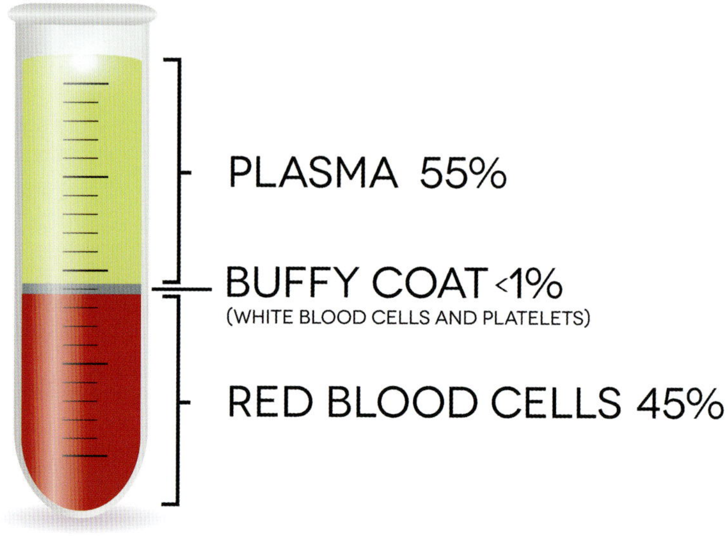

Figure 1-3 *Blood components. Hematocrit = 45%.*

A high MCH suggests that an anemia is macrocytic, where the RBCs are larger than normal (e.g., vitamin B12 or folate deficiency).

Mean Corpuscular Hemoglobin Concentration (MCHC) provides information about the average concentration of hemoglobin relative to the size of the RBC. It is a calculated value:

$$MCHC = Hgb/Hematocrit \times 100$$

While MCH measures the absolute amount of hemoglobin per RBC, MCHC gives an idea of how densely packed the hemoglobin is in the red cell.

A low MCHC is often associated with the pale (hypochromic) cells of iron deficiency anemia, where hemoglobin in the cell is less concentrated.

A high MCHC can indicate spherocytosis or dehydration, where the hemoglobin in the red cell is more densely packed.

Mnemonic: the approximate rule of 3's. *3 × RBC count = Hgb; 3 × Hgb = Hct*. For example, a normal red cell count of 5,000 corresponds to a Hgb of about 15 and a Hct of about 45. However, this only applies when the RBCs are normal. In thalassemia trait, for instance, the red cell count may be increased, while the Hgb and Hct are decreased.

Mean Corpuscular Volume (MCV)

The average volume of the red blood cell (**MCV**) is a useful criterion in determining the cause of an anemia. **Figure 1-4** lists clinical conditions that can alter the MCV.

FIGURE 1-4 MEAN CORPUSCULAR VOLUME (MCV)

INCREASED MCV (increased size of red blood cells) can be due to:
Abnormal Red Cell Production in the bone marrow

- *Megaloblastic anemia* (B12 deficiency, folic acid deficiency)
- *Heavy alcohol use* may be accompanied by a decreased intake of B12 and folic acid or may directly interfere with RBC development.
- *Hemolysis* (Red cell destruction results in increased numbers of *reticulocytes*, which are incompletely developed red cells released early from the bone marrow to deal with the hemolysis and chronic hemorrhage. Reticulocytes are larger than normal RBCs.)
- *Myelodysplasia* (a pre-leukemic bone marrow that releases larger, immature cells)
- *Drugs*. Drugs that interfere with DNA metabolism and retard normal maturation of red blood cells in the marrow (e.g., methotrexate, azathioprine) may result in a megaloblastic anemia. Certain drugs may also cause hemolysis.

Lab Artifact, e.g., cold agglutinins and rouleaux formation, where red cells clump and are counted as larger cells.

DECREASED MCV can be due to:

- *Iron deficiency anemia*, where RBCs are poorly developed, pale (hypochromic) and microcytic
- *Beta-thalassemia trait*, where RBCs are poorly developed, missing part of the hemoglobin protein, and microcytic
- *Lead poisoning* (lead may compete with iron for absorption)
- *Anemia of chronic illness* (iron, while present, is not effectively utilized to form red cells)
- *Congenital sideroblastic anemia* (defective RBC formation through failure of iron to incorporate into heme)

Red Cell Distribution Width (RDW)

RDW is a measure of the range of sizes of the red cells, which normally varies between 6-8 micrometers. **Figure 1-5** lists clinical conditions that can cause an altered RDW.

FIGURE 1-5 RED CELL DISTRIBUTION WIDTH (RDW)
INCREASED RDW (anisocytosis) can be found in: • *Iron deficiency anemia.* An anemia with a *normal RDW* may suggest thalassemia minor, which can otherwise be confused with iron deficiency anemia. • *B12 and folate deficiency.* A high RDW can provide a tipoff to a B12 or folate deficiency even before those anemias are detected. High RDW and *high* MCV suggest megaloblastic anemia (e.g., folate/B12 deficiency). High RDW and *low* MCV occur in iron-deficiency anemia. • *Sickle cell anemia* • *Myelofibrosis* (a bone marrow cancer with extensive bone marrow scarring. It disrupts normal RBC production and is associated with a severe anemia.) • *Hemolytic anemias* (reticulocytes released by the bone marrow are larger than normal RBCs with which they mix). • *A mixture of a microcytic and macroscopic anemia* in one patient • Conditions where *fragments of blood cells*, *agglutinated blood cells*, or *abnormally shaped blood cells* are counted along with normal cells • A *false high RDW* can occur if ethylenediaminetetraacetic acid (EDTA)-anticoagulated blood is used instead of citrated blood (may be due to clumping). Not every anemia has an increased RDW. **NORMAL RDW** can be found in: • *Anemia of chronic disease* • *Hereditary spherocytosis* • *Acute blood loss* (no time for reticulocytosis) • *Aplastic anemia* (bone marrow failure – including failure to produce reticulocytes) • Some cases of *thalassemia minor*

Reticulocyte Count

Increased numbers of reticulocytes appear in the blood when the bone marrow rapidly produces many new red cells due to erythropoietin action and releases immature RBC reticulocytes into the bloodstream (**Fig. 1-6**).

Figure 1-7 lists conditions that can change the reticulocyte count.

Iron

Iron is an essential ingredient of the heme in hemoglobin. **Figure 1-8** lists clinical conditions that can change the amount of serum iron.

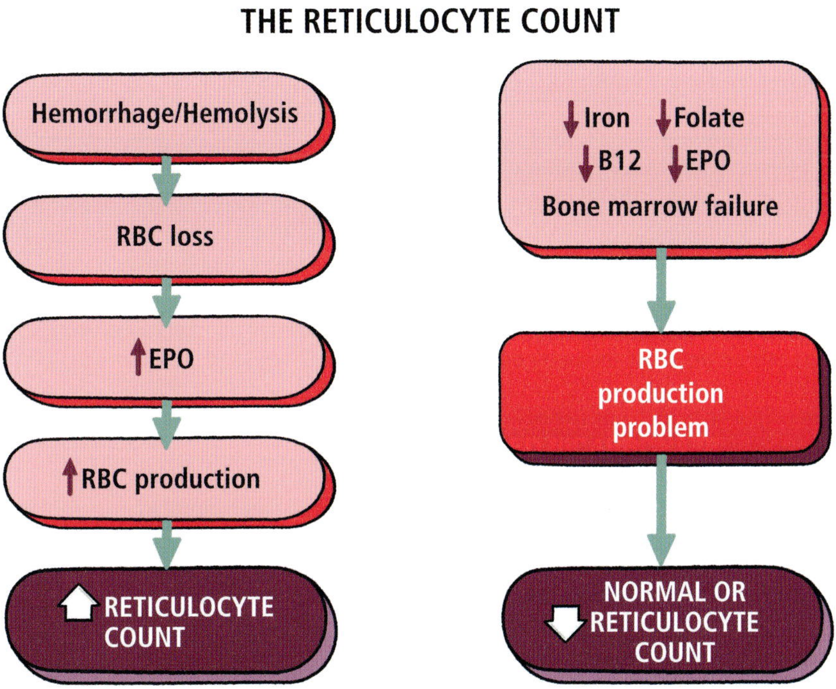

Figure 1-6 *From Berkowitz, A. Clinical Pathophysiology Made Ridiculously Simple, Medmaster.*

FIGURE 1-7 RETICULOCYTE COUNT
INCREASED RETICULOCYTE COUNT can be found in: • *Hemorrhage* (erythropoietin release) • *Response to hemolysis* • *Move to high altitude (anoxia)* due to increased erythropoietin production • *Response to treatment of anemias* (corrected with iron, B12, or folate) • *Erythropoietin-secreting tumors* (e.g., adrenal cell carcinoma, hepatocellular carcinoma, hemangioblastoma)
DECREASED RETICULOCYTE COUNT may indicate: • *Deficiency of B12, folic acid, or iron.* Reticulocyte count is low because there aren't enough ingredients to form new red cells, including reticulocytes. • *Anemia of chronic disease* (decreased production of RBCs and reticulocytes) • *Bone marrow failure* (aplastic anemia, tumor, certain drugs, radiation, infection)

FIGURE 1-8 SERUM IRON
INCREASED SERUM IRON can be due to: ***Increased iron intake*** • *Dietary* • *Iron poisoning* • *Hemochromatosis*, a disease of excess iron absorption ***Excess release of iron from cells*** • *Hemolytic anemia* • *Blood transfusions* • *Lead toxicity* (lead displaces stores of iron) • *Liver disease* (e.g., hepatitis, which releases iron stores into the blood. Iron builds up because the body does not have a good way to excrete excess iron.) **DECREASED SERUM IRON** can be due to: ***Poor dietary intake of iron*** ***Malabsorption of iron*** • P*rolonged achlorhydria* (acid is required to release iron from food) • S*prue/celiac disease* • E*xcess starch/clay eating* (may inhibit iron absorption) • S*hort bowel syndrome* (Iron is normally absorbed in the small intestinal duodenum and upper jejunum) ***Excess iron loss*** • *Hemorrhage* (e.g., GI bleeding, uterine bleeding, hematuria) • *Late pregnancy* (where the fetus depletes the mother's iron supply)

Ferritin

Iron is mostly stored in **ferritin**, a protein complex mainly stored in the liver (in hepatocytes) but also in macrophages in the bone marrow, spleen, duodenum, skeletal muscle and other areas. There is very little ferritin in the blood, but, importantly, enough to be tested. **Figure 1-9** lists clinical conditions that can cause abnormal ferritin levels.

FIGURE 1-9 SERUM FERRITIN
INCREASED SERUM FERRITIN can be due to: ***Increased iron storage need*** (which requires more ferritin to store the excess iron in an iron overload) • *Hemochromatosis* (a genetic disease of excess iron absorption) • *Hemosiderosis* (accumulation of iron from multiple transfusions or chronic hemolysis). • *Hemolysis* (red cell destruction) releases more iron to store.

- *Megaloblastic anemia.* Iron is not being incorporated into blood cells and needs to be stored.
- *Iron poisoning*
- *Hyperthyroidism* (may increase ferritin synthesis)

Inflammation (Inflammation generates an immune response that prevents iron from being utilized, but is stored instead, so more ferritin is needed to store the available iron):
- *Chronic infection*
- *Certain cancers*
- *Autoimmune diseases*
- *Kidney and liver disease*

DECREASED SERUM FERRITIN can be due to:
Decreased serum iron (since there is less iron, there is less need to produce ferritin to store it):
- *Iron deficiency anemia* (less iron requires less ferritin to store it.)
- *Dialysis* (can reduce iron during the process)

Decreased serum protein (ferritin, a protein, decreases with protein deficiency)

A low serum ferritin level is usually diagnostic of iron deficiency anemia (IDA). Ferritin is a storage form of iron and it is the *storage of iron that depletes first* before the anemia is manifest. Ferritin decline is an earlier sign of iron deficiency than other parameters of the CBC, such as red cell morphology, Hct, Hgb or serum iron (**Fig. 1-10**).

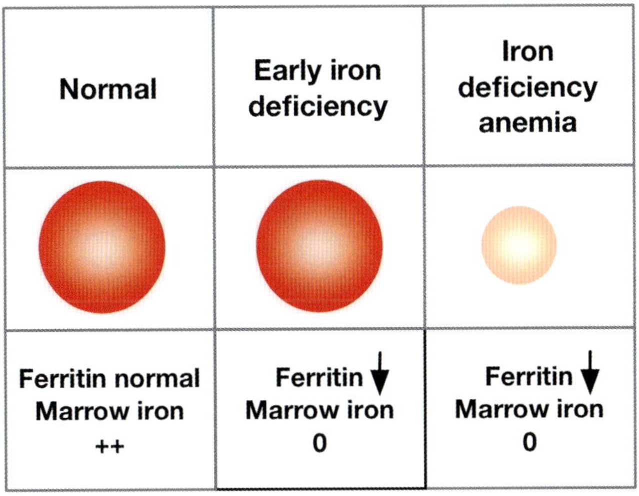

Figure 1-10 *Stages of iron deficiency. Storage forms of iron are the first to decline.*

Total Iron Binding Capacity (TIBC)

*Transf*errin is the main protein that *trans*ports iron from one place to another in the blood, for use or storage. **Total Iron Binding Capacity (TIBC)**, sometimes called **transferrin iron-binding capacity**, measures the maximum amount of iron that can be carried by the transferrin in a blood sample. The less transferrin there is, the less iron that it can bind. TIBC is an indirect measure of the amount of blood transferrin and is less expensive to test than directly measuring transferrin.

- *Iron deficiency* increases transferrin synthesis (and TIBC) in the body's effort to scrounge around for more iron in the blood for delivery to the tissues. As the amount of transferrin increases in iron deficiency anemia, the percentage of iron *saturation* in transferrin decreases.
- *Increased serum iron* (iron overload, as in hemochromatosis, where the body absorbs too much iron) reduces transferrin synthesis (and TIBC), since the body tries to decrease the overload of iron in the blood. The body tries to keep a balance in the amount of iron that is available. The less transferrin, the less iron it can bind. So TIBC decreases with iron overload (transferrin saturation increases).
- *Anemia of chronic disease* (ACD) (as in chronic infection, autoimmune diseases, cancer, kidney disease) reduces transferrin synthesis. This is a preventative way the body can reduce the availability of iron as a nutrient for bacteria and tumor cells. In ACD, iron is kept in storage and not utilized. So the ferritin increases to deal with the excess iron storage, while the transferrin (TIBC) decreases.

In general, regardless of the disease, whenever the ferritin level is up, transferrin (TIBC) level is down, and vice-versa (**Figs. 1-11** and **1-12**). If you already know the ferritin is up or down, you can predict whether the TIBC level is down or up respectively.

The diagnosis of the type of anemia is not always clear when there is a coinciding mixture of iron deficiency (e.g., from bleeding) with an anemia of chronic disease (e.g., cancer). The former lowers the ferritin, while the latter raises it.

A bone marrow biopsy will reveal absent iron staining in iron deficiency anemia, but not in anemia of chronic disease. A bone marrow biopsy, though, is invasive, and it may be best, if possible, to rely on other aspects of the history, physical, and lab to determine the diagnosis.

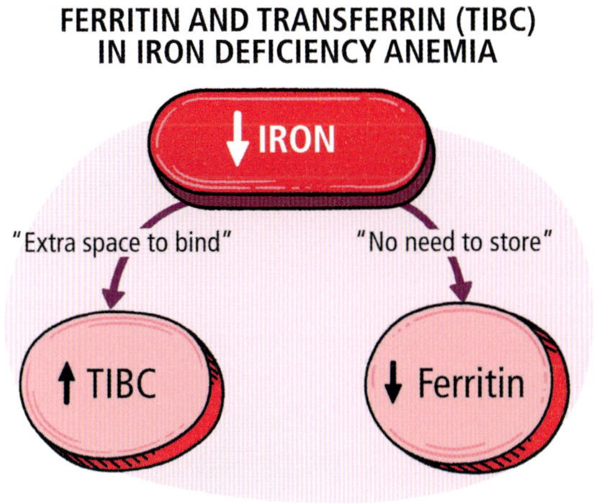

Figure 1-11 *From Berkowitz, A. Clinical Pathophysiology Made Ridiculously Simple, Medmaster.*

FERRITIN AND TRANSFERRIN (TIBC) IN IRON DEFICIENCY ANEMIA AND ANEMIA OF INFLAMMATION

	IRON DEFICIENCY ANEMIA	ANEMIA OF INFLAMMATION
FERRITIN	↓	↑
TRANSFERRIN (TIBC)	↑	↓

Figure 1-12 *From Berkowitz, A. Clinical Pathophysiology Made Ridiculously Simple, Medmaster.*

Figure 1-13 lists clinical conditions that can cause a change in TIBC.

FIGURE 1-13 TOTAL IRON BINDING CAPACITY (TIBC)
INCREASED TIBC (transferrin) can be due to: ***Iron deficiency*** (there is more transferrin, so its capacity to bind iron is high in iron deficiency anemia) ***Estrogen excess*** (pregnancy, birth control pills) leads to increased transferrin synthesis independent of body iron level.

DECREASED TIBC can be due to:

Increased serum iron (there is less need to produce more transferrin)

- *Hemolytic anemia* (iron overload)
- *Increased dietary iron intake*
- *Hemochromatosis*, a disease of excess iron absorption
- *Pernicious anemia* (iron is not incorporated into red cells; instead, it is stored in ferritin).

Decreased transferrin synthesis can be due to:

- *Cirrhosis* (transferrin is normally synthesized in the liver, which is damaged)
- *Malnutrition/hypoproteinemia* (transferrin is a protein that may be decreased in malnutrition)
- *Atransferrinemia*, a rare hereditary disease where there is an absence of transferrin

Chronic inflammation

Anemia of chronic disease (chronic infection, autoimmune diseases, cancer, kidney disease). There is poor utilization of available iron, which remains stored, coupled with decreased production of transferrin.

Transferrin Saturation

Transferrin saturation is the percentage of transferrin (transporter) protein that is saturated with iron. In general, transferrin saturation increases with iron overload and decreases with iron deficiency. Although serum iron, ferritin, and TIBC (Total Iron-Binding Capacity, also called Transferrin Iron-Binding Capacity) also give you information about the status of serum iron, transferrin saturation adds to the assessment. Serum iron alone can fluctuate throughout the day; ferritin is also an acute phase reactant, which can increase in response to inflammation or infection even if iron is low; TIBC measures the blood's capacity to bind iron with transferrin but does not give direct information about how much iron is currently being transported and can be decreased in chronic diseases, despite iron deficiency. *Low transferrin saturation usually suggests iron deficiency. High transferrin saturation commonly suggests iron overload.*

$$\text{Transferrin saturation (\%)} = (\text{Serum Iron Level} / \text{TIBC}) \times 100)$$

Figure 1-14 summarizes CBC lab values in anemia.

FIGURE 1-14 LAB VALUES IN ANEMIA (nl=normal)

	RBC	Hgb/Hct	MCV	RDW	Iron	Ferritin	TIBC	Retics	Transferrin Saturation (%)
Iron deficiency	↓	↓	↓	↑	↓	↓	↑	↓	↓
B12/folate deficiency	↓	↓	↑	↑	?↑	↑	↓	↓	nl
Anemia of chronic disease	↓	↓	↓	nl	↓	↑	↓	↓	↓
Hemolytic anemia	↓	↓	↑	↑	↑	↑	↓	↑	↑
Sickle cell	↓	↓	↑/↓	↑	↑	↑	↓	↑	↑
Thalassemia	nl/↑	↓	↓	↑	↑	↑/nl	↓	↑/nl	↑
RBC enzyme deficiency	↓	↓	↑	↑	↑	↑	↓	↑	↑/nl
Porphyria	↓	↓	nl	↑	↑	↑	↓	↑	↑/nl
Sideroblastic anemia	↓	↓	↓	↑	↑/nl	↑	↓/nl	↓	↑
Aplastic anemia	↓	↓	↑	nl	↑	↑	↓	↓	↓/nl
Hemochromatosis	nl	nl/↑	nl	(nl)	↑	↑	↓	↑	↑
Iron overload					↑	↑	↓		↑

Erythrocyte Sedimentation Rate (ESR) and C-Reactive Protein (CRP)

The **erythrocyte sedimentation rate (ESR)** measures how fast red cells settle to the bottom in an anticoagulated tube. Normal ESR is about 1-20 mm/hr, males in the lower range. ESR may be elevated in anemia, infection, pregnancy, aging, inflammation, autoimmune diseases, some kidney diseases and cancers. (High fibrinogen and other blood proteins cause red cells to stick together and settle faster.) The ESR is a rather nonspecific test that generally just tells you that "something is wrong" and may merit further investigation, particularly if it is more than 100 mm/hr. The ESR can be used to monitor the progress of the diagnosed disease.

C-reactive protein (CRP) is a plasma protein that rises in response to inflammation. While not pointing to what or where the inflammation is, its level can be used to monitor the response to medication or the progression of the disease in question. CRP may be helpful in predicting long-term prognosis and in identifying patients at risk for atherosclerotic disease (< 1 mg/dL low risk; 1-3 mg/dL moderate risk; > 3 mg/dL high risk) who might benefit from treatment with lipid lowering therapy ("statins") which lowers CRP in addition to LDL cholesterol levels.

White Blood Cells (WBCs)

White blood cells are an important part of the immune system. They help guard against infections and foreign substances.

White Blood Cell (WBC) Count

An elevated white blood cell count is common with infections, inflammation, and leukemia. In leukemia, the count can be extremely high (100,000-400,000 per microliter, in contrast with the normal range of about 4,000-11,000).

The type of white cell that increases depends largely on the kind of infection and the kind of leukemia. The white cell count may actually show a decline in overwhelming bacterial infections that use up neutrophils faster than they can be replenished, particularly in the elderly. WBC numbers may all decline in bone marrow failure, as a side effect of certain drugs (e.g., in chemotherapy), radiation, and as a response to hypersplenism (the spleen, among other things, removes white cells), cachexia, and as a congenital hereditary condition. All the white blood cells may show an artifactual rise in the setting of dehydration.

Neutrophil Count

Neutrophils characteristically predominate as a normal response to bacterial infections, although they may also predominate in certain viral and other infections (**Fig. 1-15**).

FIGURE 1-15 NEUTROPHIL COUNT
INCREASED NEUTROPHIL COUNT (neutrophilia) may occur in:
Acute infections
• *Bacterial*
- *Cocci* (staphylococcus, pneumococcus, streptococcus, gonococcus, meningococcus)
- *Bacilli* (E. coli, pseudomonas, actinomyces)
• *Certain fungi* (Coccidioides immitis)
• *Spirochetes*
• *Viruses* (chicken pox, herpes zoster, rabies, polio, smallpox)
• *Rickettsia*
• *Parasites* (liver fluke)
Inflammation/Necrosis/Trauma
• *Myocardial infarction*
• *Gout*
• *Thyroiditis*
• *Burns*
• *Postoperative period*
• *Acute glomerulonephritis*
• *Rheumatic fever*
• *Hypersensitivity reactions*
• *Collagen vascular diseases*
Myeloproliferative disorders
• *Acute myeloid leukemia*
• *Chronic myeloid leukemia*
• *Myelofibrosis* (neutrophil count may also be normal or decreased in some)
• *Myeloid metaplasia*
Metabolic disorders
• *Diabetic ketoacidosis*
• *Uremia*
• *Preeclampsia*
• *Gout*
Toxic chemicals and drugs
• Lead, mercury, digitalis, phenacetin, quinidine, turpentine
• Insect venom

Acute hemorrhage: The neutrophilia may result from inflammation and/or release of corticosteroids and epinephrine, which raise the WBC count.

***Rapidly growing tumors*:** Neutrophilia may be due to tumor necrosis factor-alpha, which induces inflammation.

Exercise may relate to epinephrine release, which raises WBC count.

Corticosteroids (e.g., Cushing syndrome): Corticosteroids cause neutrophils to detach from the endothelium and enter the bloodstream, increasing the neutrophil count. In infection, though, corticosteroids may prevent the expected rise in in WBC count.

Hereditary (***leukocyte adhesion deficiency -*** leukocytes can't leave blood vessels to migrate to the site of infection to form pus, and accumulate in the blood)

Idiopathic

DECREASED NEUTROPHIL COUNT (**neutropenia**) may result from:
Certain infections (including overwhelming bacterial infections, particularly in the elderly):

- Typhoid
- Brucellosis
- Measles
- Influenza
- Epstein-Barr virus
- Cytomegalovirus
- Viral hepatitis
- HIV-1
- Toxoplasmosis
- Tuberculosis
- Malaria
- Dengue fever
- Babesiosis
- Rickettsial infection

Bone marrow failure

- *Aplastic anemia*
- *Hematologic malignancy* (leukemia, lymphoma, myeloma, myelodysplasia)
- *Radiation treatment*
- *Metastatic tumor to bone marrow*
- Myelofibrosis
- Granulomatous infection

Immune-mediated neutropenia

- *Drug origin*
- *Autoimmune origin*
- *Rheumatoid arthritis*
- *Lupus*

- *Sjogren's syndrome*
- *Hodgkin lymphoma*
- *Autoimmune hepatitis*
- *Thymoma*
- *Goodpasture disease* (anti-neutrophil antibodies)
- *Wegener granulomatosis* (anti-neutrophil antibodies)

Hypersplenism (the spleen destroys white cells at an abnormal rate)

Cachexia

Hereditary
- *Chediak-Higashi syndrome* (immune deficiency with recurrent infections; reduced pigment in skin and eyes)
- *Cyclic neutropenia* (recurrent episodes of reduced neutrophils in blood)
- *Cartilage-hair hypoplasia syndrome* (immune deficiency with recurrent infections; skeletal abnormalities and sparse hair)
- *Dyskeratosis congenita* (bone marrow failure; rashes; white patches in mouth, misshapen fingernails)
- *Infantile genetic agranulocytosis* (*Kostmann syndrome* – lack of neutrophils; susceptibility to infections)
- *Lazy leukocyte syndrome* (neutropenia with abnormal neutrophil motility)
- *Myelokathexis* (neutrophils fail to enter the bloodstream)
- *Schwachman-Diamond syndrome* (bone marrow failure; poor growth)
- *Reticular dysgenesis* (immunodeficiency; poor WBC development; deafness)

An increased neutrophil count may occur in response to corticosteroids because

1. Steroids promote **demarginalization** of neutrophils, namely the tendency of neutrophils to detach from the blood vessel endothelium and mix with the other neutrophils floating in the blood. *Epinephrine* also promotes demarginalization. There is no actual increase in the neutrophils, but more are counted.
2. The neutrophils transmigrate less through the vessel wall and remain in the bloodstream.
3. Glucocorticoids and epinephrine induce the release of non-segmented bands of neutrophils from the blood marrow. Exercise also promotes leukocytosis, perhaps as a reaction to epinephrine release.

Lymphocyte Count

Lymphocytes tend to predominate in viral infections, but there are certain bacterial infections where lymphocytes are prominent. A rise in lymphocytes tends to appear later than neutrophils in inflammation. **Figure 1-16** lists factors that can cause a change in lymphocyte count.

Lymphopenia occurs when the body doesn't make enough lymphocytes or the lymphocytes the body makes are destroyed or sequestered in the spleen or lymph nodes.

FIGURE 1-16 LYMPHOCYTE COUNT
INCREASED LYMPHOCYTE count (lymphocytosis) may occur in: ***Viral infections*** • *Mononucleosis* • *Primary infection with HIV* • *Viral hepatitis* • *Mumps* • *Chicken pox* • *Cytomegalovirus* • *Rubella* • *Infectious lymphocytosis* • *Coxsackie virus B2* • *Poliovirus* • *Enterovirus* ***Bacterial infections*** • *Chronic bacterial infections* • *Pertussis (Bordetella pertussis)* • *Cat scratch fever (Bartonella henselae)* • *Brucellosis* • *Tuberculosis* ***Other infections*** • *Syphilis* • *Malaria* ***Hypersensitivity reactions*** ***Chronic inflammation*** ***Lymphocytic leukemia*** • *Chronic lymphocytic leukemia* • *Acute lymphocytic leukemia* ***Stress*** (lymphocytosis may be due to epinephrine, which causes release of lymphocytes from spleen, lungs, bone marrow and other lymphoid tissues; also, decreased adherence of lymphocytes to blood vessel endothelium causes more lymphocytes to remain in the blood) • *Trauma* • *Cardiac emergencies* • *Status epilepticus* ***Other*** • *Smoking* • *Post-splenectomy* (less destruction of lymphocytes) • *Radiation*

DECREASED LYMPHOCYTE COUNT (lymphopenia) may occur with:

Viral infections

- *HIV/AIDS*
- *Ebola*
- *Advanced viral hepatitis*
- *Influenza*
- *Measles*
- *SARS-CoV-2 (COVID-19)*

Bacterial infections

- *Tuberculosis*
- *Typhoid fever*
- *Sepsis*

Fungi/Parasites

- *Histoplasmosis*
- *Malaria*

Drugs

- *Chemotherapy/radiotherapy*
- *Immunosuppressants* (steroids)

Autoimmune disorders

- *Rheumatoid arthritis*
- *Lupus*
- *Aplastic anemia*
- *Myasthenia gravis*
- *Chronic kidney disease*
- *Sarcoidosis*
- *Multiple sclerosis*

Hematologic cancer and other blood diseases

- *Leukemias and metastatic tumors of the bone marrow*
- *Aplastic anemia*

Hereditary

- *Hematopoietic stem cell aplasia*
- *Ataxia-telangiectasia* (immunodeficiency; decreased coordination; spider veins)
- *Cartilage-hair hypoplasia syndrome* (immune deficiency with recurrent infections; skeletal abnormalities and sparse hair)
- *Idiopathic CD4+ lymphocytopenia*
- *Thymoma immunodeficiency*
- *Severe combined immunodeficiency syndrome*
- *Wiskott-Aldrich syndrome* (immune deficiency; eczema; low platelet count)
- *Adenosine deaminase deficiency* (immune deficiency; infections)
- *Purine nucleoside phosphorylase deficiency* (immune deficiency; infections; neurologic symptoms; autoimmune disease)
- *DiGeorge anomaly* (immunodeficiency; congenital anomalies)

Malnutrition

Stress

Eosinophil Count

Eosinophils increase in allergies, in certain autoimmune diseases, and in response to parasitic worm infections. Eosinophils are prominent in a variety of tumors. **Figure 1-17** lists factors that can cause a change in eosinophil count.

Decreased eosinophil count does not usually indicate a problem unless all the other white cell counts are low, too.

FIGURE 1-17 EOSINOPHIL COUNT

INCREASED EOSINOPHIL COUNT may occur in:

Allergy
- *Asthma*
- *Allergies*
- *Autoimmune disorders* (atopic dermatitis, Crohn's disease, ulcerative colitis, and other autoimmune diseases)

Parasitic worm infections
- *Ascariasis*
- *Schistosomiasis*
- *Trichinosis*
- *Visceral larva migrans*
- *Strongyloidiasis* (roundworm)
- *Gnathostomiasis* (roundworm)
- *Fascioliasis* (liver fluke)
- *Paragonimiasis* (lung fluke)

Other infections
- *Scarlet fever* (group A strep)
- *Epstein-Barr virus*
- *Aspergillus*
- *HIV*
- *Tuberculosis*

Tumors
- *Gastric carcinoma*
- *Lung carcinoma*
- *Lymphoma*
- *Hodgkin lymphoma*
- *Non-Hodgkin lymphoma*
- *Human T-cell lymphotropic virus (HTLV-1)*
- *Adult T-cell leukemia/lymphoma (ATLL)*
- *Eosinophilic leukemia*

Idiopathic

DECREASED EOSINOPHIL COUNT may occur in:
- *Alcohol intoxication*
- *Excess cortisol production (e.g., Cushing syndrome) or Rx with corticosteroids*
- *Morning* as opposed to evening

Monocyte Count

Monocytes are prominent in certain infections as well as certain autoimmune diseases, cancers and other conditions.

Figure 1-18 lists conditions that can change the monocyte count.

FIGURE 1-18 MONOCYTE COUNT

INCREASED BLOOD MONOCYTES can be found in:

Bacterial infections
- *Tuberculosis*
- *Brucellosis*
- *Subacute bacterial endocarditis*

Spirochete infection
- Syphilis

Viral infection
- Infectious mononucleosis

Protozoan and rickettsial infections
- Malaria
- Rocky Mountain spotted fever
- Kala-azar

Neoplasia
- *Hodgkin disease*
- *Monocytic leukemia*
- *Myeloproliferative diseases*

Autoimmune diseases
- Lupus
- *Rheumatoid arthritis*
- *Ulcerative colitis* and other inflammatory bowel diseases

Other
- *Sarcoid*
- *Lipid storage diseases*
- *Dehydration* (artifactually elevates all leukocytes due to hemoconcentration)

DECREASED BLOOD MONOCYTES

A low monocyte count can be due to anything that decreases the overall white cell count, with accompanying neutropenia and lymphopenia.

Basophil Count

Basophils increase in certain infections and inflammations, leukemias, hemolytic anemias, following splenectomy, and certain endocrine disorders.

Figure 1-19 lists factors that can cause a change in basophil count.

FIGURE 1-19 BASOPHIL COUNT

INCREASED BASOPHIL COUNT may be found in:

Myeloproliferative disorders

- *Chronic myeloid leukemia*
- *Polycythemia vera*
- *Primary myelofibrosis*
- *Essential thrombocythemia*
- *Myelodysplastic syndrome*
- *Systemic mastocytosis*
- *Hypereosinophilic syndrome*

Inflammation

- *Inflammatory bowel disease*
- *Rheumatoid arthritis*
- *Asthma*
- *Chronic sinusitis*
- *Chronic dermatitis/psoriasis*
- *Hashimoto thyroiditis*

Allergies

- *Food allergies*
- *Drug allergies*
- *Hay fever*
- *Allergic rhinitis*

Infections

- *Chicken pox*
- *Tuberculosis*
- *Parasitic infections*

Endocrine

- *Hypothyroidism*
- *Increased estrogen*

DECREASED BASOPHILS may be found with:

- *Hypersensitivity reactions* (e.g., allergic urticaria)
- *Hyperthyroidism*
- *Corticosteroids*
- *Lupus*
- *Pregnancy*
- *Hereditary absence of basophils*

Platelets

Platelets are critical elements in the initiation of blood clotting.

Platelet Count

Figure 1-20 lists factors that can cause a change in platelet count.

FIGURE 1-20 PLATELET COUNT
INCREASED PLATELET COUNT may be found in: ***Excess production of platelets*** • *Essential thrombocytosis* • *Acute hemorrhage* • *Acute infection* • *Arthritis and other chronic inflammations* • *Chronic myelogenous leukemia* • *Lymphoma* • *Certain tumors of the colon* • *Polycythemia vera* • M*yeloid metaplasia* • *Iron-deficiency anemia* (iron deficiency induces a general attempt by the marrow to produce cells. Platelet production occurs, since it is not dependent on the presence of iron.) • *Osteoporosis* • *Stress* • *Surgery* ***Decreased elimination of platelets*** • *Post-splenectomy* (the spleen normally helps remove platelets) ***Normal physiological difference*** **DECREASED PLATELET COUNT** may occur with ***Defective production of platelets*** • *Leukemia* and other myelofibrosis diseases (The marrow is replaced, including the capacity to produce platelets) • *Cancer chemotherapy* (destroys bone marrow) • *Pernicious anemia* (B12 is needed for platelet production) • *Wiskott-Aldrich syndrome* (immune deficiency; bleeding; eczema) • *Bernard-Soulier disease* (platelet defect; bleeding) ***Increased removal of platelets*** • *Hypersplenism* • *Hemorrhage* (platelets are lost during bleeding) • *Immune thrombocytopenic purpura (ITP)* • *Cirrhosis with hypersplenism* (the spleen removes platelets) • *Posttransfusion purpura* (immune-mediated platelet destruction)

- *Neonatal alloimmune thrombocytopenia* (maternal antibodies against fetal platelet antigens). Healthy women may have an inconsequential mild thrombocytopenia during pregnancy.
- *Drug-induced thrombocytopenia* (antibiotics, cardiac drugs, anti-rheumatics, heparin)
- D*isseminated intravascular coagulation (DIC)* (platelets are used up) *Infection* (can sometimes result in thrombocytopenia from immune destruction or DIC)
- *Thrombotic thrombocytopenic purpura* (There is a large amount of abnormal von Willebrand protein, which is involved in platelet adhesion)
- *Hemolytic uremic syndrome* (kidney failure; low red cells and platelets)
- *Artifact of platelet clumping in EDTA* (as opposed to sodium citrate) test tube anticoagulation.

Atypical Cells

Apart from being able to view differences in cell size and number, the blood smear can also assess atypical-looking cells that provide clues to the diagnosis (**Fig. 1-21** –red cell morphologies; **Fig. 1-22** –white cell morphologies)

Normal red blood cell (Fig. 1-21A). Non-nucleated; central pale area; without inclusions.

Microcytic hypochromic red blood cell (Fig. 1-21B). Small, pale red blood cell most commonly found in iron deficiency.

Target cells (Fig. 1-21C) (increased cell surface/Hgb ratio due to either excess cell surface area or decreased hemoglobin content). The RBCs do not carry oxygen efficiently. Seen in thalassemia, iron deficiency anemia, sickle cell disease, hemoglobin C and D diseases, and after splenectomy (altered RBCs are no longer removed by the spleen). Liver disease may alter the lipid synthesis needed for the cell membrane structure.

Nucleated RBCs (Fig. 1-21D) in severe hemolysis with premature release of the altered form. Normally, 1% or less of RBCs should be nucleated.

Basophilic stippling (Fig. 1-21E). Numerous basophilic granules of aggregated RNA, ribosomes, and mitochondria; indicates disturbed erythropoiesis (e.g., thalassemia, hemolytic anemia, sickle-cell, myelodysplastic syndrome, leukemia, and heavy metal poisoning, such as lead).

Reticulocytes (Fig. 1-21F). Immature non-nucleated red cells released early from the bone marrow to deal with hemolysis and chronic hemorrhage. Reticulocytes are larger than normal RBCs. Staining shows residual RNA and mitochondria.

ATYPICAL RED BLOOD CELLS

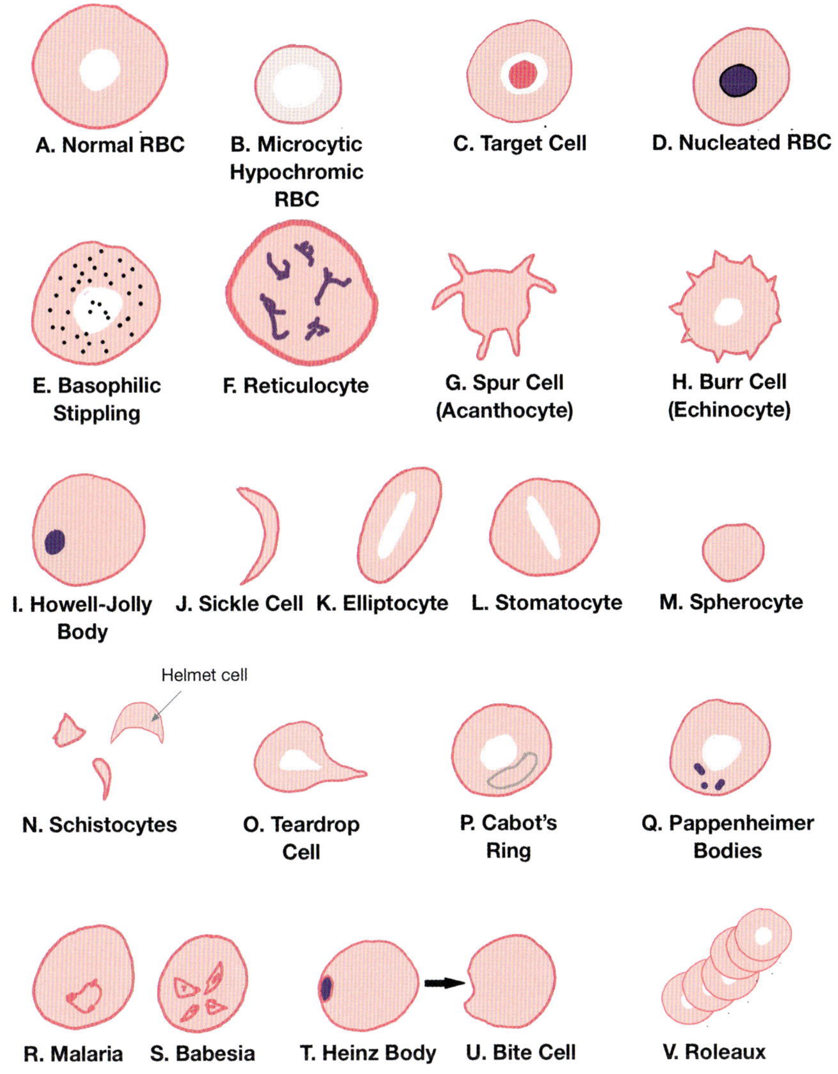

Figure 1-21

Spur cells (acanthocytes) (Fig. 1-21G) RBCs with thorny projections in cirrhosis, pancreatitis and other conditions. Cells have poor deformability. They arise from membrane defects in protein or lipids (e.g., apolipoprotein A-II deficiency). Seen in *abetalipoproteinemia*. Not like Burr cells where spikes are shorter and at regular intervals on the cell membrane.

Burr cells (echinocytes) (Fig. 1-21H) in kidney failure (uremia) and in hemolytic anemia – short evenly spaced spikes like sea urchins. May also be an artifact of anticoagulation. Spikes are shorter and more numerous than in spur cells. Mechanism: outer lipid bilayer expands outward relative to the inner lipid bilayer.

Howell-Jolly bodies (Fig. 1-21I). Little fragments of RBC nucleus that the spleen normally removes. May be seen in any condition where there is loss of splenic function.

Sickle cells (Fig. 1-21J) in sickle cell anemia.

Elliptocytes (Fig. 1-21K) in hereditary elliptocytosis (problem with membrane protein), iron deficiency anemia, thalassemia, megaloblastic anemia, myelodysplastic syndrome.

Stomatocytes (Fig. 1-21L). *Stomatocytosis* is a rare condition where the red cells have a slit-like central zone instead of the usual round central pallor. The cell is fragile and subject to hemolysis. The condition may be hereditary or the result of excess alcohol ingestion, in which case it may disappear after about 2 weeks of alcohol withdrawal.

Spherocytes (Fig. 1-21M). Loss of membrane surface proteins and surface area produce a spherical shape. Found in hereditary spherocytosis and immunologically caused *extravascular hemolysis* (destruction of blood cells outside the blood vessels, in the spleen, liver, and bone marrow).

Schistocytes (Fig. 1-21N). Jagged fragments (triangular, **helmet** or comma-shaped), of RBCs seen in *intravascular hemolysis*. Those cells with a few very large protuberances are called **horn cells**.

Teardrop cells (dacryocytes) (Fig. 1-21O) are found in bone marrow fibrosis and myeloproliferative diseases. May be caused by mechanical squeezing out from the bone marrow as a result of bone marrow infiltration.

Cabot's rings (Fig. 1-21P) are rare threadlike looping strands possibly of microtubules found in cases of abnormal RBC production in B12 deficiency and lead poisoning.

Pappenheimer bodies (Fig. 1-21Q) are cytoplasmic iron granules found in sideroblastic anemia.

Parasites (Figs. 1-21R, S). Malaria (Fig. 1-21R), a single celled parasitic organism, presents as irregular inclusions on Giemsa stain, often ring-shaped. **Babesia (Fig. 1-21S)** is a parasite spread by a tick bite. Symptoms may resemble those in malaria. Its configuration in some of the red cells resembles a Maltese cross.

Heinz bodies (Fig. 1-21T*)* are denatured clumped hemoglobin (seen in G6PD deficiency and other RBC enzyme defects). When spleen macrophages bite them out, they are then called **bite cells (Fig. 1-21U).**

Rouleaux formation (Fig. 1-21V) (like a stack of coins) in connective tissue diseases, diabetes, multiple myeloma (immunoglobulin-coated RBCs cling together) antibiotic allergy, and diabetic retinopathy. Occurs where plasma protein (particularly fibrinogen and globulins) is high.

Atypical White Blood Cells

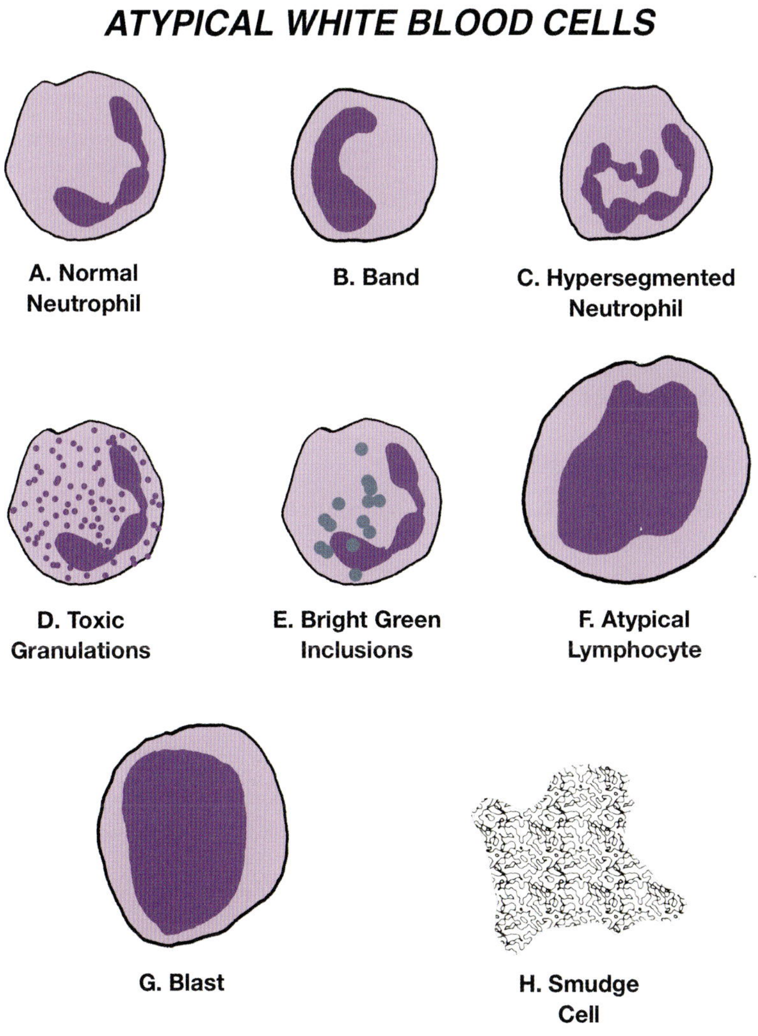

Figure 1-22

Normal neutrophil (Fig. 1-22A). The nucleus contains 3-5 segments.

Bands (Fig. 1-22B) – immature neutrophils in serious infections (>15% bands = "left shift")

Hypersegmented neutrophils (Fig. 1-22C) (6 or more lobes or > 3% of neutrophils with at least 5 nuclear lobes) in B12 and folate deficiency, and myeloproliferative disorders. Reason unclear. **Bilobed neutrophils** are found in *Pelger-Huet anomaly,* a rare inherited benign condition.

Toxic granulation in neutrophils (Fig. 1-22D). Large granules in the cytoplasm of segmented and band neutrophils in blood are found in severe infections. Indicates phagocytosis and lysosomal activity.

Bright green inclusions (Fig. 1-22E) in liver failure (lipid inclusions) (poor prognosis)

Atypical (reactive) lymphocytes (Fig. 1-22F) are large, distorted lymphocytes with large irregularly shaped nuclei. Found in certain infections, particularly viral (e.g., mononucleosis, cytomegalovirus, hepatitis B), immune reactions to transplantation and immunization, autoimmune diseases, cancer, and drug reactions.

Blasts (Fig. 1-22G). Immature WBCs (**myeloblasts, metamyelocytes, promyelocytes, myelocytes**, or in the case of the lymphatic system, **lymphoblasts** or **prolymphocytes**). Suspicious for leukemia.

Smudge cells (basket cells) (Fig. 1-22H) are characteristic of chronic lymphocytic leukemia. They are the remnants of abnormally fragile, ruptured lymphocytes and are without nuclear or cytoplasmic structure.

Atypical Platelets

Platelets too large (increased mean platelet volume, MPV) are seen in *immune thrombocytopenic purpura, Bernard-Soulier disease,* and *white platelet syndrome,* an autosomal dominant disorder with thrombocytopenia and prolonged bleeding.

Platelets too small (decreased MPV) are found in thrombocytopenia arising from an aplastic anemia, as well as in *Wiskott-Aldrich syndrome* (an immune deficiency syndrome, with susceptibility to infections and bleeding).

Hypogranular platelets are found in *white platelet syndrome* (thrombocytopenia; increased bleeding tendency; enlarged platelet size, abnormal internal platelet structure; decreased responsiveness to stimuli to platelet aggregation).

2

Bleeding and Clotting Tests

Platelet Function

Platelets are fragments of large megakaryocyte cells in the bone marrow. Despite their small size (about 3 micrometers), platelets perform a number of tasks important to the clotting process. A defect in any of these functions will lead to the risk of bleeding (**Fig. 2-1**)

Platelets initiate blood clotting with a platelet plug (**Fig. 2-1**):

1. Platelets stick to the damaged blood vessel wall, thereby participating in platelet clot formation. **Von Willebrand factor (vWF)**, a glycoprotein, is the initial bridge that connects platelets to the injured vessel wall. **GP Ib-IX** is a protein complex on the platelet cell surface that binds to von Willebrand factor.
2. Platelets contract, due to actin and myosin molecules within them, thereby facilitating clot retraction and closure of the damaged blood vessel.
3. Platelets secrete ADP, thromboxane A2, and serotonin, which activate other platelets to become sticky and join in the clot-forming process.
4. Platelet IIb/IIIa protein increases the binding of platelets to fibrinogen and facilitates platelets sticking together.

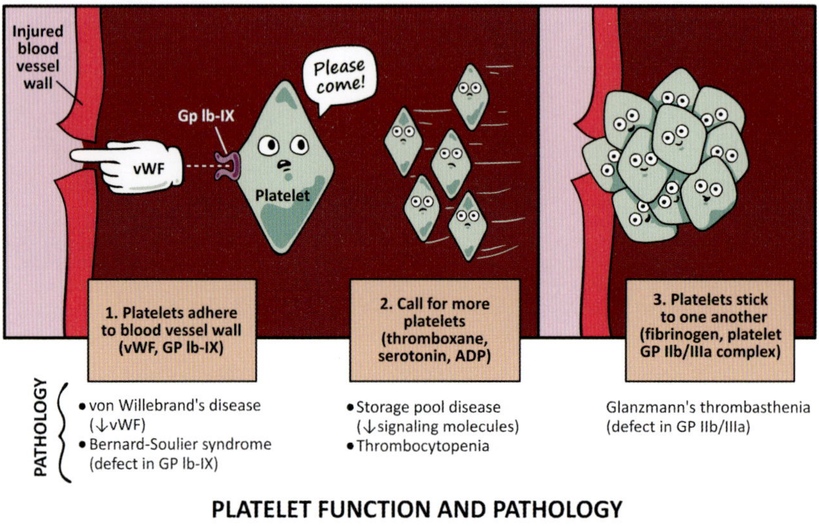

Figure 2-1 *Platelets. (Modified from Berkowitz, A. Clinical Pathophysiology Made Ridiculously Simple, Medmaster)*

5. Platelet phospholipids and platelet factor 3 (platelet tissue factor) initiate thrombin formation and are important ingredients in the cascading pathway of blood clotting.
6. Platelets produce fibrin-stabilizing factor (**Factor XIII**), which binds fibrin molecules into a meshwork and strengthens the clot. **Factor XIII deficiency**, a rare autosomal recessive disease, impairs the stability of the clot, which breaks down, with a resulting bleeding tendency.
7. Platelets produce prostaglandins, which have a number of different effects on blood flow and wound healing.

The Blood-Clotting Cascade

When a blood vessel is severed, a series of processes aid in wound closure:

1. Local constriction of the blood vessel, due to local neurogenic causes and local tissue chemical reactions, including platelet release of serotonin
2. Formation of a pure platelet plug, in the case of small wounds to capillaries or very small blood vessels. The pure platelet plug is often sufficient to close small wounds, even without a blood clot.

3. Formation of a blood clot, in the case of larger wounds. A blood clot contains not only platelets, but red blood cells and the end products of the blood-clotting processes, particularly fibrin.
4. Wound repair with formation of fibrous tissue

Blood clotting involves a number of clotting factors (**Fig. 2-2**):

Factor I (Fibrinogen)
Factor II (Prothrombin)
Factor III (Tissue thromboplastin)
Factor IV (Calcium)
Factor V (Proaccelerin)
Factor VII (Proconvertin)
Factor VIII (Antihemophilic factor A). Von Willebrand factor is the carrier for Factor VIII.
Factor IX (Antihemophilic factor B)
Factor X (Stuart factor)
Factor XI (Antihemophilic factor C)
Factor XII (Hageman factor)

High molecular weight kininogen (Fitzgerald factor) and **prekallikrein (Fletcher factor)** accelerate the *early* steps in intrinsic path clotting at the level of Factor XII.

Factor XIII (fibrin stabilizing factor) acts at the *end* of the clotting cascade by linking together fibrin molecules and strengthening the clot.

Absence of any factor may lead to the failure of clotting.

Blood clotting, apart from the platelet plug includes a sequence of steps (**Fig. 2-2**):

1. Injury to the blood vessel wall or to the blood cells themselves induces, after a number of initial steps, the formation **of prothrombin activator.**
2. **Prothrombin activator** catalyzes the change of **prothrombin** to **thrombin.**
3. **Thrombin** changes **fibrinogen** to **fibrin** threads, which mix with RBCs, platelets, and plasma to form the blood clot.
4. **Clot retraction**, assisted by platelets, expresses serum (serum is plasma minus its clotting factors, such as fibrinogen, etc. which are used up in forming the clot).

Sounds relatively simple if you start with prothrombin activator. But what are the steps that lead to the formation of prothrombin activator? That is more complex. There are an extrinsic pathway and an intrinsic pathway, both of which lead to the formation of the prothrombin activator (**Fig. 2-2**).

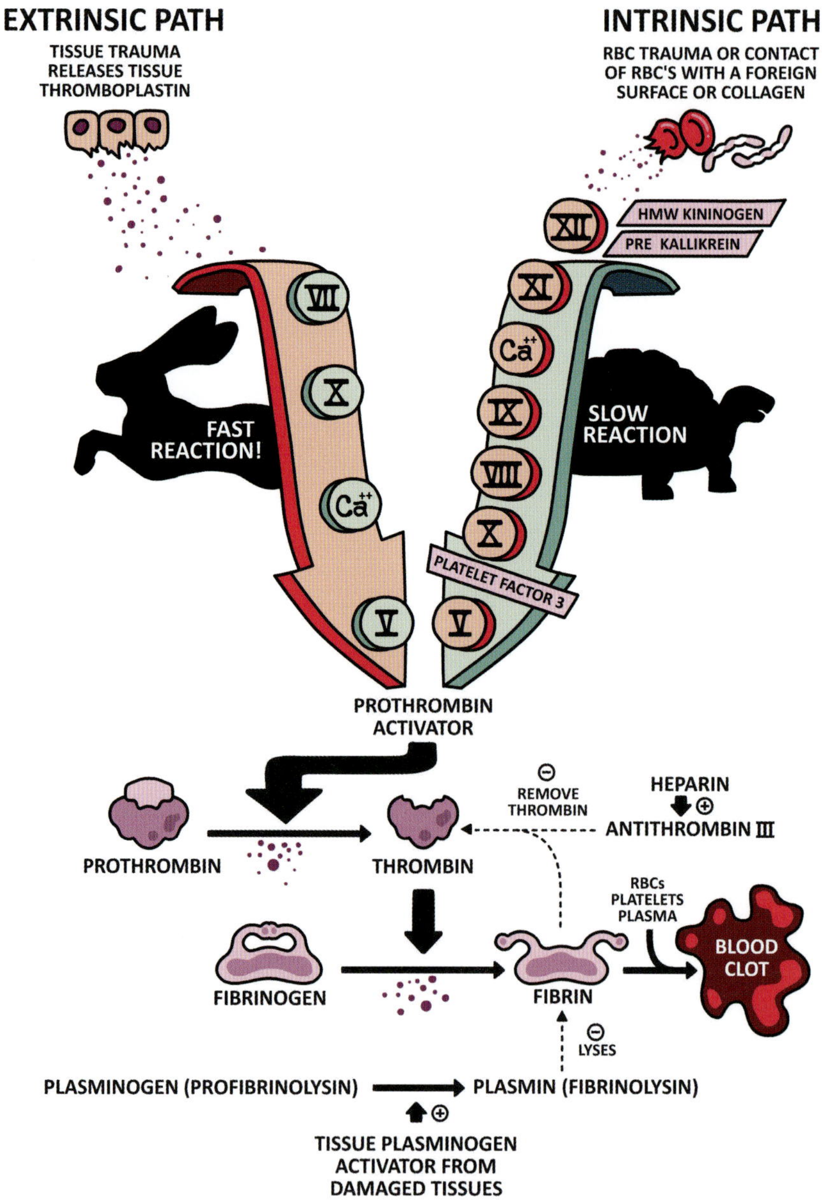

Figure 2-2 *Clotting pathways. (From Goldberg, S. Clinical Physiology Made Ridiculously Simple, Medmaster)*

The **extrinsic pathway** is triggered by tissue **thromboplastin**, which is released from damaged cells in the vascular endothelial wall or outside the blood vessel.

The **intrinsic pathway** is triggered by damage to the blood cells themselves, or by contact of the blood cells with a foreign surface, such as skin or collagen in the wound, or the glass in a blood collection tube. Hence, unless a collected blood specimen is mixed with an anticoagulant, it will clot in the glass tube due to activation of the intrinsic pathway.

Both the intrinsic and extrinsic pathways involve clotting factor enzymes that participate in cascading reactions that ultimately lead to the formation of the prothrombin activator.

Figure 2-2 shows a simplified schema of the ingredients of the intrinsic and extrinsic cascading pathways of blood clotting. The extrinsic pathway normally proceeds quickly; blood may clot within 11-15 seconds of activation of the extrinsic pathway. Clotting in vivo may take several minutes, however, along the intrinsic pathway, which has more steps than the extrinsic pathway.

Coagulation Tests

Tests for the integrity of the coagulation mechanism are based on whether you want to test the intrinsic or extrinsic pathway, or platelet function, as follows:

1. **Tests of the intrinsic pathway**: The **whole blood clotting time test** measures the time taken for blood to clot in a glass test tube (normally about 2-8 minutes). This tests the intrinsic pathway, which is activated by contact with a foreign surface, such as glass. If any component of the intrinsic pathway is defective, the clotting time will be prolonged over normal controls. You can then identify the component that is defective by assaying for particular clotting factors.

More accurate evaluation of the intrinsic and extrinsic pathways may be done using the partial thromboplastin time (PTT) and the prothrombin time (PT).

The PTT evaluates the intrinsic pathway.

The PT evaluates the extrinsic pathway.

The type of tube used to draw blood depends on whether you want to examine whole anticoagulated blood or coagulated blood serum (**Fig. 2-3**).

The **partial thromboplastin time (PTT)**, like the whole blood-clotting time, tests for the integrity of the intrinsic pathway. Clotting in a test tube is initially prevented by citrating the plasma to remove calcium, a necessary

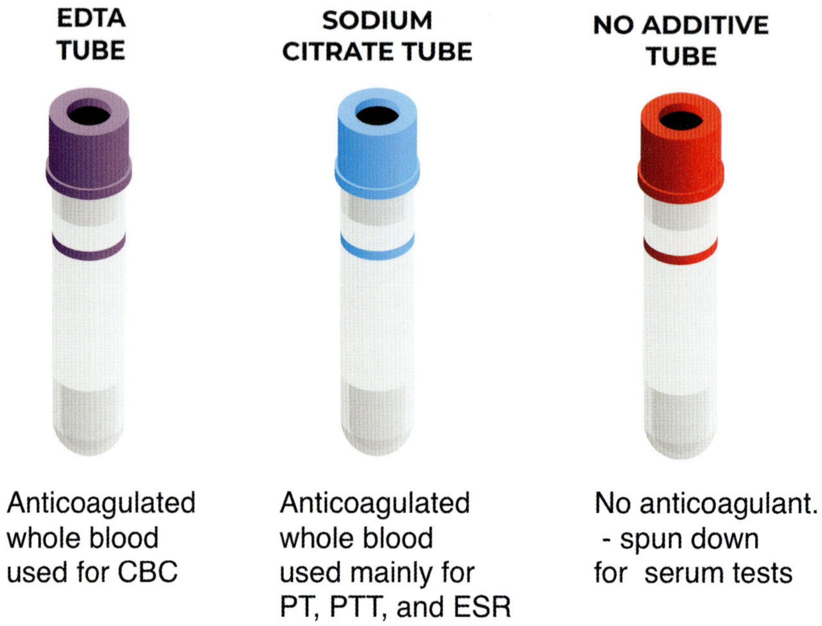

Figure 2-3 *Blood collection tubes.*

molecule in the extrinsic and intrinsic pathways. The PTT measures the time taken for recalcified citrated plasma to clot in the test tube, usually about 60-70 seconds, depending on the laboratory. The **aPTT (activated partial thromboplastin time)** is a more sensitive variation on the PTT in which an activator is added to the plasma sample to speed up the clotting process, commonly about 25-35 seconds. The aPTT is preferred for monitoring heparin therapy.

2. **Tests of the extrinsic pathway**: The **one-step prothrombin time (PT)** is the time needed for recalcified citrated plasma to clot in the *presence of tissue thromboplastin*. This "protime" test adds the critical tissue ingredient (tissue thromboplastin) that is necessary to start off the relatively fast extrinsic pathway. The normal protime is about 11–15 seconds, depending on the lab. If the protime is prolonged, in relation to controls, this suggests a problem somewhere in the extrinsic pathway.

Mnemonic: The extra "T" that is "intrinsic" to "PTT" stands for thromboplastin.

FIGURE 2-4 PROTHROMBIN TIME (PT) AND PARTIAL THROMBOPLASTIN TIME (PTT) TESTS		
	PT	**PTT**
Intrinsic path defect	Normal	Abnormal
Extrinsic path defect	Abnormal	Normal
Liver failure, vitamin K deficiency, coumarin or heparin therapy	Abnormal	Abnormal

The **INR** is a way of measuring clotting time that is more accurate than the prothrombin time. While the PT is a relatively simple and inexpensive test, it varies considerably depending on the laboratory and in itself is not a reliable way to determine how well a patient is being anticoagulated on heparin. The INR is more reliable since it is a ratio of the patient's prothrombin time to a series of controls that have established clotting times and known INR values. A normal INR is 1.1 or less, while an INR range of 2.0 to 3.0 is considered an effective range for people taking warfarin for anticoagulation.

Neither the intrinsic nor extrinsic pathway can cause clotting if there is a defect at the end steps of prothrombin-to-thrombin or fibrinogen-to-fibrin. Therefore, *both* the prothrombin time and PTT will be abnormal in severe liver disease (the liver manufactures prothrombin, fibrinogen and other clotting factors), vitamin K deficiency (vitamin K is necessary for formation of prothrombin and other factors), coumarin therapy (coumarin interferes with the formation of the reduced form of vitamin K, and hence the formation of prothrombin and other clotting factors), and heparin administration (heparin indirectly inactivates thrombin) (**Fig. 2-4**).

While PT and PTT assess for abnormality in the extrinsic or intrinsic paths, the **thrombin time (TT)** (also called **Thrombin Clotting Time**) tests the final step of the coagulation cascade, the conversion of fibrinogen to fibrin. It tests the time for a clot to form in the plasma of a blood sample after the addition of thrombin. An increased TT indicates a delay in the conversion of fibrinogen to fibrin, which may occur in fibrinogen deficiency or the presence of thrombin inhibitors, like heparin. It is also possible to directly test for fibrinogen levels.

D-dimer is a protein fragment from clot breakdown. An elevated D-dimer suggests that there has been significant clot formation in the body, with subsequent breakdown of the clot. A normal D-dimer level can be helpful in "ruling out" deep vein thrombosis (DVT) and/or a pulmonary embolism.

3. **Tests of platelet function**. Platelet function may be diminished either because of decreased numbers of platelets or because of a deficiency in the functioning of existing platelets.

A platelet count (about 150,000-450,000 platelets/microliter in adults) is, of course, useful in assessing the number of platelets. Bleeding may occur with low platelet counts. The **bleeding time** assesses platelet function. A small cut is placed on the patient's forearm, with a blood pressure cuff kept on the arm at 40mm Hg to resist venous flow. Normal bleeding time (depending on method) may be 2–8 min. Bleeding time tends to be normal in coagulation disorders of the extrinsic and intrinsic pathway, because the platelet plug operates independently of these pathways and is sufficient by itself to close such small wounds. If the bleeding time is prolonged, this usually suggests a defect in platelet function.

It is also possible to test for deficiency of individual components of the clotting process, such as antithrombin and von Willebrand factor.

Figures 2-5 and **2-6** list the causes of an increased or decreased prothrombin time.

FIGURE 2-5 CAUSES OF INCREASED PROTHROMBIN TIME (PT)

Liver disease (the liver produces clotting factors)
- Cirrhosis
- *Hepatitis*
- *Liver failure*

Vitamin K deficiency (vitamin K is needed to synthesize several clotting proteins)
- *Inadequate dietary intake of vitamin K*
- *Malabsorption disorders* (e.g., celiac disease, Crohn's disease)
- *Long-term antibiotic use* that affects gut flora

Anticoagulant medications
- Warfarin (Coumadin)
- Direct oral anticoagulants

***Disseminated Intravascular Coagulation* (DIC):** A condition where small blood clots develop throughout the bloodstream, consuming clotting factors and leading to prolonged PT

Inherited clotting factor deficiencies
- Hemophilia
- Von Willebrand disease

***Lupus anticoagulant*:** An antibody against phospholipids that increases activated partial thromboplastin time (aPTT) by interfering with phospholipid-dependent steps in the intrinsic coagulation pathway

Other blood disorders
- *Leukemia* (can infiltrate the liver)
- *Myelodysplastic syndromes* (may affect liver function)

Medications and supplements that affect liver or vitamin K function
- *Antibiotics* (e.g., cephalosporins)
- *Antifungals* (e.g., fluconazole)
- *High doses of vitamin E* (can interfere with vitamin K metabolism)

Malabsorption syndromes: Conditions that impair nutrient absorption can lead to deficiencies in clotting factors.

Chronic kidney disease: Severe renal impairment can interfere with vitamin K absorption and with liver function.

Consumption coagulopathy: Conditions that lead to excessive consumption of clotting factors (e.g., massive bleeding, extensive surgery)

FIGURE 2-6 CAUSES OF DECREASED PROTHROMBIN TIME (PT)

Vitamin K excess: Excessive intake of vitamin K through diet or supplements can decrease PT.

Hypercoagulable states: Conditions like Factor V Leiden mutation or prothrombin gene mutation can make blood more prone to clotting.

Hormonal influences: Pregnancy or oral contraceptive use can lead to a hypercoagulable state, reducing PT.

Thrombophilia: Inherited or acquired disorders that increase the risk of blood clot formation

Dehydration: Severe dehydration can concentrate blood and decrease PT.

Polycythemia: An increased number of red blood cells can make the blood more viscous and reduce PT.

Acute phase reaction: Inflammatory states can lead to increased levels of clotting factors, reducing PT.

3

The Comprehensive Metabolic Panel

The comprehensive panel provides information about blood sugar levels, electrolytes, and kidney and liver function.

Glucose

Glucose levels are important in diagnosing and managing diabetes, evaluating for excess blood glucose (**hyperglycemia**) or deficiency in blood glucose (**hypoglycemia**).

Figures 3-1 and **3-2** list the potential causes of hyperglycemia and hypoglycemia.

Hemoglobin A1c

Excess glucose normally enters red blood cells and links up (glycates) with hemoglobin. This glycated hemoglobin (HbA1c) is relatively slow to build up and slow to disappear and can be measured as an index of average blood glucose over the past 2-3 months. Even if the current glucose is normal, an HbA1c level ≥ 6.5% suggest diabetes (prediabetes: 5.7%-6.4%). HbA1c is usually measured every 3-6 months. It provides a monitor of the effectiveness of glucose control and risk for diabetic complications, such as retinopathy, nephropathy, neuropathy, and cardiovascular disease.

FIGURE 3-1 CAUSES OF HYPERGLYCEMIA

INCREASED INTAKE OF GLUCOSE

- *Increased dietary intake of carbohydrates*
- *Increased IV intake of glucose*

EXCESS PRODUCTION OF GLUCOSE

Secondary to increased cortisol

- A*cute stress response*: Causes increased cortisol production, which stimulates **gluconeogenesis** (synthesis of glucose from non-carbohydrate sources, such as amino acids, lactate, and glycerol)
- *Cushing's syndrome* (iatrogenic or otherwise): Excess cortisol stimulates gluconeogenesis.

Secondary to increased glucagon

- G*lucagonoma*: Glucagon increases **glycogenolysis** (breakdown of glycogen to glucose) and gluconeogenesis.
- *Acute pancreatitis*: Glucagon leaks out of inflamed pancreatic cells.

Secondary to increased catecholamines

- *Pheochromocytoma*: There is excess catecholamine production; catecholamines increase glycogenolysis (breakdown of glycogen to glucose).

DECREASED REMOVAL OF GLUCOSE

Secondary to decreased insulin production: Insulin is needed to drive glucose into cells. Insulin also inhibits gluconeogenesis and promotes the storage of glucose as glycogen. Lack of insulin in type I diabetes or late type II diabetes causes hyperglycemia.

Secondary to insulin resistance

- *Type II diabetes mellitus:* Insulin is present but cells are resistant to it.
- *Chronic renal failure*: The uremic patient may develop insulin resistance, with decreased glucose uptake by the muscles.
- *Diuretics*: Diuretics induce insulin resistance.
- *Growth hormone-secreting tumor* (acromegaly): Promotes resistance at the level of the insulin receptor and also stimulates glucagon
- *Pregnancy* (hormone changes in pregnancy cause insulin resistance)
- *Beta-blockers*: Beta-blockers can inhibit pancreatic insulin release, as well as decrease peripheral sensitivity to insulin.

FIGURE 3-2 CAUSES OF HYPOGLYCEMIA

DECREASED GLUCOSE INTAKE

Starvation

DECREASED GLUCOSE PRODUCTION

- *Hypothyroidism*: Thyroid hormone stimulates gluconeogenesis and glycogenolysis.
- *Adrenal insufficiency*: There is cortisol deficiency. Cortisol promotes gluconeogenesis and inhibits peripheral utilization of glucose. Deficiency in cortisol promotes hypoglycemia.
- *Liver disease*: The liver stores glycogen and changes it to glucose. This process may be deficient in liver disease.
- *Alcohol*: Alcohol metabolism in the liver produces excess NADH, which can reduce gluconeogenesis.
- *Glycogen storage diseases*: Can impair glycogenolysis and gluconeogenesis
- *Unripe ackee fruit (Jamaican Vomiting Sickness***)**: The unripe fruit contains a chemical (hypoglycin) that inhibits gluconeogenesis.

EXCESS GLUCOSE REMOVAL

Strenuous exercise (uses up glucose)

Secondary to increased insulin

- *Insulin or other overdose of drugs that decrease glucose*
- *Insulinoma*
- *Hypopituitarism*: Growth hormone has anti-insulin-like activity. Lack of growth hormone promotes hypoglycemia.
- *Kidney disease*: Can impair the clearance of insulin
  Congenital hyperinsulinism

Electrolytes

Sodium (Na+)

Sodium is critically important in maintaining fluid balance; nerve, muscle, and other cellular function; blood pressure regulation; acid-base balance; and many aspects of cell function. Sodium (Na+) is the most common cation in the blood. Chloride (Cl-) is the most common anion followed by bicarbonate (HCO3-). Measuring sodium levels helps assess fluid and electrolyte balance, hydration, kidney function, and certain metabolic conditions, like Addison's disease or Syndrome of Inappropriate Antidiuretic Hormone secretion (SIADH).

Figures 3-3 and **3-4** list the potential causes of too much sodium in the blood (**hypernatremia**) and too little sodium (**hyponatremia**).

FIGURE 3-3 CAUSES OF HYPERNATREMIA

INCREASED INTAKE OF SODIUM

- *Increased dietary intake of sodium*
- *Increased IV fluid intake of sodium (e.g., hypertonic saline solutions, sodium bicarbonate)*

DECREASED EXCRETION OF SODIUM

- *Hyperaldosteronism*: Sodium is then reabsorbed by the kidney rather than excreted.
- *High renin disorders*: Stimulate aldosterone secretion and sodium retention
- *Cushing's syndrome*: Corticosteroids have some aldosterone-like effect.
- *Conn's syndrome (primary hyperaldosteronism)*: The adrenals produce excess aldosterone.

INADEQUATE INTAKE OF WATER

- *Primary hypodipsia:* Sodium then becomes concentrated.

INCREASED WATER LOSS (more so than sodium loss)

Skin

- *Sweating*
- Burns

Respiratory

- *Fever*
- *Dry environment*
- *Hyperventilation*

Renal

- *Osmotic diuretics:* Cause increased water excretion
- *Glycosuria:* Glucose excretion drags water along.

Diabetes insipidus: Decreased antidiuretic hormone (ADH) results in excess renal water excretion.

- *Central diabetes insipidus*: A deficiency of ADH of pituitary or hypothalamic origin
- *Nephrogenic diabetes insipidus*: Renal unresponsiveness to ADH

Gastrointestinal

- *Diarrhea*
- *Vomiting*

FIGURE 3-4 CAUSES OF HYPONATREMIA

DECREASED INTAKE OF SODIUM

- *"Tea and toast" diet* (typically low in sodium)
- *Excess beer drinking* (beer is low in sodium)
- *Decreased Na+ in IV administration*

INCREASED EXCRETION OF SODIUM

Dehydration (when there is more sodium loss than water loss, especially when more water is ingested to replace the fluid loss)

- *Diarrhea*
- *Vomiting*
- *Sweating*
- *GI tube drainage*
- *Diuretics*
- *Nasogastric suction*
- *Intestinal ileus*
- *Hypoaldosteronism*: Aldosterone is needed for Na+ retention by the kidney.
- *Cortisol deficiency*: Cortisol has some aldosterone-like effect.
- *Salt-wasting nephropathy*

EXCESS INTAKE OF WATER

- *Excess oral water intake*
- *Excess IV water intake*

DECREASED EXCRETION OF WATER

- *Syndrome of Inappropriate ADH secretion (SIADH)*: Reabsorption of too much water by the kidney
- *Congestive heart failure*: Decreased glomerular filtration rate (GFR) results in decreased water excretion.
- *Cirrhosis*: Dilutional hyponatremia. Ascites fluid pools in abdomen; the kidneys think the body is dehydrated and hold onto more water.
- *Hypothyroidism*: Decreased cardiac output and decreased GFR result in decreased water excretion.
- *Adrenal insufficiency*: Reduced aldosterone and cortisol lead to impaired sodium reabsorption.

PSEUDOHYPONATREMIA: Low concentration of sodium is due to presence of other serum osmoles in excess (e.g., glucose), which draw water into the blood stream, thus diluting the sodium.

- *Severe hypertriglyceridemia*
- *Severe hyperproteinemia*
- *Hyperglycemia*

Potassium (K+)

Potassium is important in nerve and muscle function, fluid balance, blood pressure regulation, pH balance, and many other aspects of metabolism. Potassium levels help diagnose **hyperkalemia** (excess potassium concentration in blood) or **hypokalemia** (low potassium concentration in the blood) and are used to follow patients on diuretics or certain heart medications.

Figures 3-5 and **3-6** list the causes of hyperkalemia and hypokalemia.

FIGURE 3-5 CAUSES OF HYPERKALEMIA

INCREASED POTASSIUM (K+) INTAKE

- *Increased K+ in diet*
- *Increased K+ in IV*

DECREASED K+ REMOVAL

- *Renal failure*: The kidney can't clear K+.
- *Addison's disease*: Lack of aldosterone results in defective K+ excretion.
- *Dehydration*
- *Aldosterone-inhibiting diuretics*
- *ACE (angiotensin-converting enzyme) inhibitors*: ACE promotes the synthesis and release of aldosterone, which causes sodium retention and potassium excretion. ACE inhibitors cause sodium loss and potassium retention.
- *Beta-blockers*: May decrease uptake of K+ into cells. (Catecholamines increase uptake of K+ into cells.)
- *Hyporeninemic hypoaldosteronism:* Lack of renin decreases aldosterone synthesis and release.
- *Tubular unresponsiveness to aldosterone*
- *Diabetes mellitus* (insulin deficiency or resistance): Insulin normally moves K+ into cells.

RELEASE OF K+ IN BLOOD

Clinical hemolysis

- *Hemolytic anemias*
- *Hemolysis of transfused blood*

Redistribution h***yperkalemia*** (potassium leaks out of cells):

- *Acidosis*: Excess blood H+ enters cells in exchange for K+.
- *Tissue injury* (K+ leakage)
- *Infection*
- *Massive digitalis overdose* (digitalis causes cells to release K+)
- *Hyperkalemic periodic paralysis* (a condition in which muscle cell K+ channels open, allowing K+ to exit into the circulation)

PSEUDOHYPERKALEMIA

Lab artifact (Pseudohyperkalemia): A blood sample is damaged during or after blood drawing, allowing K+ to leak out of the sample cells. The patient's actual K+, though, is normal:

- *Marked thrombocytosis (Platelets > 1,000,000):* K+ leaks out of the platelets in the blood sample.
- *Severe leukocytosis (WBC > 200,000):* The white blood cells in the sample leak K+.
- *Familial pseudohyperkalemia*: A genetic mutation causes K+ leaks from RBCs in blood samples.
- *Mononucleosis* (white cell fragility in a blood sample causes artifactual K+ leakage)

FIGURE 3-6 CAUSES OF HYPOKALEMIA

DECREASED POTASSIUM INTAKE

- *Decreased K+ in diet*
- *Decreased K+ in IV*

INCREASED POTASSIUM LOSS

Gastrointestinal

- *Diarrhea*
- *Laxative abuse*
- *Vomiting*
- *Villous adenoma of large bowel*: Can cause excess mucus and electrolyte secretion, including K+
- *Malabsorption of potassium (e.g., celiac disease, Crohn's disease)*

Renal

- *Diuretics* (loop and thiazide diuretics)
- *Hyperaldosteronism.* Aldosterone causes sodium reabsorption and potassium loss.
- *Cushing's syndrome*: Corticosteroids have some aldosterone-like effect.
- *Burns*: K+ loss is partly due to increased aldosterone secretion.
- *Trauma*: K+ loss is partly due to increased aldosterone secretion.
- *Type I (distal) renal tubular acidosis:* Increased K+ secretion
- *Type II (proximal) renal tubular acidosis:* Increased K+ secretion
- *Osmotic diuresis:* High urinary flow can "wash out" K+.
- *Saline administration:* Increases extracellular fluid volume and dilutes K+; increased urine output, with K+ loss)
- *Renal artery stenosis*: Aldosterone, which causes Na+ retention and K+ loss, is secreted because of decreased renal blood flow.
- *Bartter's syndrome*: Defect in the kidney's ability to reabsorb K+
- *Gitelman's syndrome*: Defect in the kidney's ability to reabsorb K+, along with hypomagnesemia

Skin
- Sweating

Endocrine
- *Diabetic ketoacidosis*: Osmotic diuresis is accompanied by potassium loss.
- *Liddle syndrome* (a genetic syndrome with increased Na+ reabsorption and K+ excretion)
- *Hyperaldosteronism* (causes Na+ reabsorption and K+ excretion)
- *Cushing syndrome*: Cortisol has an aldosterone-like effect.

Drugs
- *Carbonic acid anhydrase inhibitors* (diuretics that result in net K+ excretion)
- *Aminoglycoside antibiotics*: Toxic effect on distal tubule
- *Cis platinum*: Toxic effect on distal tubule
- *Sodium penicillins*: Can have diuretic effect with K+ loss
- *Licorice*: Has an aldosterone-like effect, reabsorbing Na+ and excreting K+

Other
- *Magnesium depletion* (magnesium deficiency interferes with the channels in the renal tubules needed for K+ reabsorption
- *Ascites*: Abdominal fluid collection results in less renal blood flow, resulting in aldosterone secretion.

REDISTRIBUTION
- *Excess insulin administration*: Insulin drives potassium into cells.
- *Excess glucose administration*: Glucose stimulates insulin secretion; insulin drives both glucose and K+ into cells.
- *Alkalosis*: In response, H+ leaves cells in exchange for K+ to reduce the alkalosis.
- *Familial hypokalemic periodic paralysis*
- *Hypokalemic periodic paralysis with thyrotoxicosis*
- *Increased beta2-adrenergic activity*: Increases the activity of Na+/K+ ATPase and drives K+ into cells
- *Theophylline toxicity*: Causes beta-adrenergic stimulation, which increases the activity of the Na+/K+ ATPase pump and shifts potassium intracellularly.
- *Factor replacement (B12, folate) in megaloblastic anemia*: Causes potassium shift into newly forming RBCs

ARTIFACT
- *Marked leukocytosis >100,000:* White cells may take up K+ in the blood specimen.
- *Insulin*: Taken just before blood drawing can move K+ into cells in the test tube.

Chloride (Cl-)

Chloride levels are used in the evaluation of electrolyte balance and acid-base disorders. **Figures 3-7** and **3-8** list the causes of too much chloride (hyperchloremia) and too little chloride (hypochloremia).

FIGURE 3-7 CAUSES OF HYPERCHLOREMIA
DEHYDRATION: Chloride ions become more concentrated. **CONDITIONS THAT CAUSE HYPERNATREMIA**: Chloride levels tend to follow sodium levels to maintain electroneutrality. (See *Hypernatremia* and *Hyponatremia*) **CONDITIONS THAT DECREASE BICARBONATE**: There is an inverse relationship between bicarbonate and chloride. If bicarbonate decreases, chloride (the most common extracellular ion) increases to maintain electroneutrality. (See *HCO3- Decreased* page) • *Metabolic acidosis*: HCO3- is low. • *Respiratory alkalosis*: HCO3- is low. • *Hyperparathyroidism*: Parathyroid hormone inhibits proximal renal tubular absorption of HCO3-. The kidneys compensate for the HCO3- loss by increasing Cl- reabsorption.

FIGURE 3-8 CAUSES OF HYPOCHLOREMIA
EXCESS WATER • *Overhydration*: Dilutional • *Syndrome of Inappropriate ADH secretion*: Water is retained by the kidney, diluting the Cl-. • *Congestive heart failure*: Water is retained in excess of sodium. **CONDITIONS THAT CAUSE HYPONATREMIA**: Chloride tends to follow sodium to maintain electroneutrality. (See *Hyponatremia* page) **EXCESS CHLORIDE LOSS** • *Vomiting*: Chloride normally is high in the stomach and GI tract due to HCl production in the stomach. • *Gastric suction* • *Diarrhea* **CONDITIONS THAT INCREASE BICARBONATE**: There is an inverse relationship between bicarbonate and chloride. If bicarbonate increases, chloride decreases to maintain electroneutrality. • *Respiratory acidosis*: HCO3- is high. Cl- decreases to maintain electroneutrality • *Metabolic alkalosis*: HCO3- is high. Cl- decreases to maintain electroneutrality

Carbon Dioxide (CO_2) and Bicarbonate (HCO_3^-)

Testing levels of CO_2 or HCO_3^- helps assess acid-base balance and respiratory or metabolic acidosis/alkalosis. Levels of CO_2 tend to go in the same direction as levels of HCO_3^-, since $CO_2 + H_2O \leftrightarrow H^+ + HCO_3^-$. I.e., more CO_2 causes more HCO_3^-. Less CO_2 causes less HCO_3^-. The kidneys regulate bicarbonate levels, while the lungs regulate CO_2 levels. While bicarbonate levels provide indirect information about CO_2, direct measurement of CO_2 with blood gas analysis is more accurate. For instance, in acute respiratory disorders, CO_2 may change quickly, but bicarbonate levels may take time to rise as the kidneys try to compensate.

Figures 3-9 and **3-10** list the causes of increased and decreased bicarbonate.

FIGURE 3-9 CAUSES OF INCREASED BICARBONATE
LOSS OF H+ IONS
• *Metabolic alkalosis*: $H_2CO_3 \rightarrow HCO_3^- + H^+$ (H^+ is lost, leaving HCO_3^-)
• *Vomiting:* Loss of stomach acid leads to loss of H^+ and an increase in blood HCO_3^-. Also, the loss of chloride leads to the retention of bicarbonate to maintain electrical neutrality.
• *Gastric suction*
• *Aldosteronism*: Stimulates Na^+ reabsorption in the renal tubule collecting duct. This leaves an electronegativity in the tubule lumen, which is corrected by excretion of H^+ (and K^+) into the lumen, with retention of HCO_3^-.
• *Loop and thiazide diuretics:* Can cause the kidneys to excrete more H^+ ions
• *Hypokalemia*: K^+ leaves cells to compensate, with H^+ responding by entering the cells, leaving an alkalosis with increased HCO_3^-.
GAIN OF HCO_3^-
• *Renal compensation for respiratory acidosis:* The kidneys reabsorb more HCO_3^- to neutralize the excess acidity in the blood.
• *HCO_3^- administration* (IV or orally, e.g., too much antacids)
LOSS OF CHLORIDE
• *In significant sweating or cystic fibrosis* (can cause bicarbonate retention to maintain electrical neutrality)

FIGURE 3-10 CAUSES OF DECREASED BICARBONATE
GAIN OF H+ IONS
• *Metabolic acidosis*: $H^+ + HCO_3^- \rightarrow H_2CO_3$ (HCO_3^- is neutralized by acid; HCO_3^- levels drop,)
• *Diabetic ketoacidosis*
• *Lactic acidosis*

- *Ingestion of acidic compounds:* Neutralizes HCO3-
- *Renal failure* (impaired bicarbonate reabsorption; reduced H+ excretion neutralizes the blood bicarbonate by the retained H+)
- *Starvation*: Ketoacids build up, which are neutralized by HCO3-.
- *Shock*: Results in lactic acidosis

LOSS OF HCO3-

- *Renal compensation for respiratory alkalosis*: Bicarbonate is a base. Excretion of bicarbonate helps correct the blood alkalosis.
- *Diarrhea* (loss of HCO3- in the GI tract)
- *Medications that decrease bicarbonate reabsorption in the kidneys* (e.g., carbonic anhydrase inhibitors)

Calcium (Ca++)

Calcium is critical for bone health; nerve, skeletal, and cardiac muscle function; blood clotting; parathyroid, and many cellular functions. Calcium levels help evaluate the health of these systems, as well as the effectiveness of treatment for hypercalcemia and hypocalcemia.

Figures 3-11 and **3-12** list the causes of hypercalcemia and hypocalcemia.

FIGURE 3-11 CAUSES OF HYPERCALCEMIA

INCREASED ABSORPTION OF CALCIUM

Vitamin D excess (vitamin D promotes intestinal absorption of Ca++):

- *Excess vitamin D ingestion*
- *Tuberculosis*: Associated with increased vitamin D production
- *Sarcoidosis*: Associated with increased vitamin D production induced by sarcoid macrophages
- *Acromegaly*: Associated with increased vitamin D production
- *Milk-alkali syndrome*: Excess ingestion of Ca++ and alkali, often from calcium-rich antacids. Also, the alkalosis stimulates Ca++ reabsorption in the distal renal tubule.

Hyperparathyroidism: Parathyroid hormone stimulates intestinal absorption of Ca++.

Paraneoplastic syndrome: Through the secretion of parathyroid hormone-related protein (PTHrP), which mimics the action of parathyroid hormone

DECREASED EXCRETION OF CALCIUM

Renal failure: Hypercalcemia occurs in some cases when renal failure is combined with increased Ca++ intake.

Hyperparathyroidism: PTH stimulates reabsorption of Ca++ in the kidney, so less calcium is excreted.

Milk-alkali syndrome: Metabolic alkalosis increases hypercalcemia by stimulating Ca++ reabsorption in the distal tubule.

Thiazide diuretics: Decrease urinary calcium excretion. Thiazides inhibit sodium reabsorption in the kidney and indirectly increase calcium reabsorption.

Addison's disease: Aldosterone deficiency causes volume depletion with resulting increased calcium concentration. There is also a deficiency of cortisol; cortisol normally inhibits calcium absorption in the intestines and promotes calcium excretion through the kidneys. Cortisol deficiency allows more calcium to enter and remain in the blood.

Lithium: Can promote hyperparathyroidism with increased reabsorption of calcium in the kidneys

INCREASED RESORPTION + DECREASED EXCRETION OF CALCIUM

Familial hypocalciuric hypercalcemia: An underactive Ca++ sensing mechanism on parathyroid cells and renal tubule cells interprets Ca++ as being too low. This results in excess PTH release and decreased calcium excretion.

BONE RESORPTION (BONE BREAKDOWN)

Hyperparathyroidism: Parathyroid hormone stimulates bone breakdown with release of calcium.

Non-parathyroid PTH-secreting tumor

Tumor metastasis to bone

Paget's disease of bone

Prolonged bedrest: Calcium then moves from bone to blood.

Hyperthyroidism: Stimulates bone resorption with release of calcium

Addison's disease: Increased bone resorption from reduced cortisol levels

FIGURE 3-12 CAUSES OF HYPOCALCEMIA

DECREASED ABSORPTION OF CALCIUM

Vitamin D deficiency: Vitamin D normally promotes calcium absorption in the intestines and kidneys.

- *Decreased intake of vitamin D*
- *Malabsorption of vitamin D*
- *Decreased activation of vitamin D* in chronic kidney disease or severe liver disease, since both the kidney and liver are important in the activation of vitamin D.

INCREASED EXCRETION OF CALCIUM

Renal failure: Calcium loss may occur because of decreased activation by the kidney of vitamin D. Vitamin D normally helps promote reabsorption of calcium by the kidney.

Hypoparathyroidism: Parathyroid hormone (PTH) increases Ca++ reabsorption in the kidney. PTH deficiency results in renal loss of Ca++.

Pseudohypoparathyroidism: Due to the resistance of the body to parathyroid hormone

Hyperphosphatemia: Phosphate binds Ca++.

Fanconi syndrome: Proximal tubule dysfunction leads to increased excretion of calcium (among other ions).

DECREASED RESORPTION + INCREASED EXCRETION OF CALCIUM

Hypercalciuric hypocalcemia: An overactive Ca++ sensing mechanism on parathyroid cells and renal tubule cells interpret Ca++ as being too high. This results in decreased PTH release and increased calcium excretion.

REDISTRIBUTION OF CALCIUM

Hypoparathyroidism: PTH promotes bone breakdown, releasing calcium. Deficient PTH reduces blood calcium.

Osteomalacia: There is decreased bone calcification, hence less Ca++ to supply to the blood.

Pancreatitis: Release of lipase leads to fat necrosis and formation of fatty acids, which react with Ca++ and reduce serum Ca++.

Fat embolism: Fat reacts with Ca++ and reduces serum Ca++.

Alkalosis: Ca++ enters cells in response to high pH in exchange for H+. Alkalosis also increases the ability of albumin to bind Ca++ to serum albumin, reducing the concentration of free, ionized calcium in the blood.

Parental magnesium: Given in pregnant women with marked preeclampsia or eclampsia to reduce neuromuscular excitability. Magnesium can suppress PTH release and cause resistance to PTH.

Hungry bone syndrome: Bones sequester calcium post-surgical correction of hyperparathyroidism.

MISCELLANEOUS

Drugs

- *Chemotherapy agents* may impair renal function and also release phosphate from cancer cells, which binds to Ca++.
- *Foscarnet*, an antiviral agent, can chelate Ca++ ions, reducing free calcium in the blood.
- *Phenytoin*, an anticonvulsant, impairs vitamin D activation in the liver.

Hypoalbuminemia: Albumin binds calcium, so decreased albumin leads to decreased *total* calcium ("fictitious hypocalcemia") even though the *free* (ionized. unbound) calcium may remain normal. Routine testing is for total (including bound) calcium.

Severe chronic illness: Can reduce Ca++ in different ways, whether by malnutrition and vitamin D deficiency; kidney disease; decreased synthesis in a damaged liver of albumin, which is necessary to bind Ca++; loss of protein in inflammatory bowel disease; and malabsorption of Ca+ in the intestines.

Dilutional hypoalbuminemia: From overhydration or fluid retention

Osteomalacia is a condition marked by decreased bone *calcification*. **Osteoporosis** is a multifactorial condition characterized by decreased bone *density and mass*.

Bone density testing with dual-energy X-ray absorptiometry (DEXA) scans can help assess changes in bone density over time. A T-score (compares the patient's bone density with that of a young, healthy adult of the same sex) of -1.0 or above is considered normal; T-scores between -1.0 and -2.5 indicate osteopenia (low bone density), while T-scores of -2.5 or below indicate osteoporosis. Conventional x-rays are unreliable for detecting changes in bone mass, since 30% of bone must be lost before the changes show on x-ray.

Magnesium (Mg++)

Magnesium is an essential body mineral, not just for electrolyte balance, but for enzyme activity; energy production; protein synthesis; bone health; and muscle, nervous system, and cardiovascular function. An excess of magnesium can cause symptoms ranging from lethargy and weakness to muscle paralysis, respiratory failure, heart block, cardiac arrest and coma. A deficiency of magnesium can also affect these systems, with symptoms of muscle weakness and spasms; cardiac arrhythmias; hypertension, confusion, and other signs.

Figures 3-13 and **3-14** list the potential causes of hypermagnesemia and hypomagnesemia.

FIGURE 3-13 CAUSES OF HYPERMAGNESEMIA

TOO MUCH MAGNESIUM ADMINISTERED

Drugs

- *Magnesium-containing antacids*
- *Magnesium-containing laxatives*: E.g., milk of magnesia
- *Vitamins*
- *Parenteral (intravenous) magnesium:* Given in pregnant women with marked preeclampsia or eclampsia to reduce neuromuscular excitability. Given in asthma to reduce airway constriction. Given for torsade de pointes or other cardiac arrhythmias to reduce abnormal electrical activity.
- *Epsom salts (magnesium sulfate):* When ingested
- *Near-drowning in the Dead Sea*: Water there is very high in magnesium

TOO LITTLE MAGNESIUM EXCRETED

Decreased kidney excretion

- *Renal failure*
- *Lithium intoxication*: Perhaps by decreasing urinary excretion of Mg++
- *Hypoparathyroidism*: Decreased calcium; calcium normally competes with renal Mg++ absorption. Too little calcium increases Mg++ absorption.
- *Decreased GI excretion*: In hypomotility disorders, which may result in increased absorption of Mg++

(Drugs that decrease GI motility; constipation; bowel obstruction)

- *Familial hypocalciuric hypercalcemia*: Low renal excretion of both calcium and magnesium

MAGNESIUM RELEASE BY CELLS

Tumor lysis syndrome: Large amounts of intracellular magnesium are released.

Rhabdomyolysis: Releases intracellular magnesium, as in tumor lysis syndrome

OTHER

Hypothyroidism: Slowed excretion of magnesium

Milk-alkali syndrome: Kidney damage reduces the excretion of magnesium.

Adrenal insufficiency: Decreased aldosterone production. Aldosterone normally promotes the excretion of both potassium and magnesium.

Dehydration: Increased Mg++ concentration

FIGURE 3-14 CAUSES OF HYPOMAGNESEMIA

DECREASED INTAKE OF MAGNESIUM

- *Poor nutrition*
- *Chronic alcoholism* (poor dietary intake, GI malabsorption)

INCREASED LOSS OF MAGNESIUM

Gastrointestinal

- *Diarrhea/vomiting*
- *Laxative abuse*
- *Regional enteritis*
- *Ulcerative colitis*
- *Malabsorption diseases*
- *Small intestine bypass surgery*
- *Pancreatitis*: Fat malabsorption; the unabsorbed fat locks onto intestinal Mg++
- *Aging:* Impaired absorption; reduced bone magnesium

Renal

- *Diuretics*
- *Nephrotoxic drugs* (aminoglycosides; amphotericin B; cisplatin; cyclosporine)
- *Hyperaldosteronism*: Aldosterone increases renal excretion of magnesium.

Familial disorders of magnesium excretion

- *Associated with hypercalciuria:* Excess urine Ca++ interferes with Mg++ reabsorption.
- *Gitelman syndrome:* A mutation impairs the ability of the distal renal tubule to reabsorb Mg++.
- *Bartter syndrome:* Impaired renal reabsorption of Mg ++ (as well as impaired reabsorption of potassium, sodium, chloride, and calcium)

Aging

Hypercalcemia: Calcium competes with renal magnesium reabsorption

Metabolic acidosis: Ca++ leaves cells in exchange for high H+; hypercalcemia causes hypomagnesemia by competing with Mg++ for renal absorption

Excess lactation: Increased demand for Mg++, a key component of milk

Marked sweating: Sweat contains magnesium.

REDISTRIBUTION OF MAGNESIUM

- *Insulin therapy*: Insulin moves Mg++ into cells.
- *Hungry bone syndrome*: Mg++ enters bone following parathyroidectomy or thyroidectomy.
- *Hyperadrenergic states*: Catecholamines move Mg++ into cells and also increase renal excretion of Mg++.

Phosphate (PO4)

Phosphate is essential for many body functions, including energy production (a component of ATP); bone and tooth structure; phospholipids in cell membranes; part of DNA and RNA; blood buffering; enzyme activation; and cell signaling. Phosphate imbalance cause numerous problems, including muscle and bone weakness, fatigue; and cardiovascular, neurologic, and digestive issues,

Figures 3-15 and **3-16** list the causes of hyperphosphatemia and hypophosphatemia.

FIGURE 3-15 CAUSES OF HYPERPHOSPHATEMIA

INCREASED INTAKE OF PHOSPHATE

- *Increased oral intake of PO4*
- *Laxatives and enemas that contain PO4*
- *Increased IV administration of PO4*
- *Vitamin D intoxication*: Vitamin D promotes intestinal absorption of PO4.
- *Sarcoidosis*: Vitamin D produced by granulomas

DECREASED LOSS OF PHOSPHATE
- *Renal failure*
- *Hypoparathyroidism*: PTH normally stimulates PO4 excretion.
- *Pseudohypoparathyroidism*: Failure of kidney to respond to PTH
- *Acromegaly*: Growth hormone promotes retention of PO4.
- *Familial tumor calcinosis*: May be a defect in the proximal renal tubule that results in increased PO4 reabsorption

REDISTRIBUTION OF PHOSPHATE
- *Bone metastasis*: Releases PO4
- *Hemolysis*: PO4 is released from destroyed RBCs.
- *Tumors*: Intracellular phosphate leaks from damaged cells (e.g., multiple myeloma, lymphoma).
- *Rhabdomyolysis*: Intracellular phosphate leaks from damaged cells.
- *Malignant hyperthermia*: Muscle cell damage results in PO4 leaking out.
- *Acidosis*: Decreased pH (Increased H+) results in PO4 leaving cells to act as a buffer (diabetic ketoacidosis; lactic acidosis: respiratory acidosis)

ARTIFACT
- *Pseudohyperphosphatemia*: An artifact of lab testing in certain conditions, such as hemolysis, hyperlipidemia, hyperproteinemia, hyperbilirubinemia and certain medications

FIGURE 3-16 CAUSES OF HYPOPHOSPHATEMIA

DECREASED PO4 INTAKE
- *Dietary deficiency of PO4*
- *Vitamin D deficiency*
- *Antacids with aluminum or magnesium: Bind PO4*
- *Malabsorption*
- *Chronic alcohol abuse:* Interferes with phosphate absorption

INCREASED PO4 EXCRETION
- *Hyperparathyroidism*: PTH increases urinary excretion of PO4
- *Vitamin D deficiency*: Renal tubules then do not reabsorb PO4
- *Fanconi syndrome:* A renal tubule disorder with phosphate wasting
- *Diabetic ketoacidosis:* Urinary loss of PO4 during diuresis

REDISTRIBUTION OF PO4
- *Lymphoma*: Consumes phosphate
- *Excess insulin*: Insulin moves PO4 into cells.
- *Respiratory alkalosis*: Moves phosphate into cells
- *Sepsis and inflammation*: Increased cellular uptake of phosphate

GENETIC
- X-linked hypophosphatemia
- Autosomal dominant hypophosphatemic rickets

Anion Gap

Sometimes, screening tests include a calculation of the **anion gap**, which mainly helps in determining the cause of metabolic acidosis by identifying *unmeasured* anions (negative ions) that accumulate in the blood. This can indicate conditions like lactic acidosis (lactate), keotoacidosis (ketone bodies), renal failure (e.g. sulfate, phosphate, urate), or toxin ingestion (e.g., formate from methanol; glycolate and oxalate from ethylene glycol; salicylate from aspirin).

Kidney Function Tests

Blood Urea Nitrogen (BUN)

BUN testing helps evaluate the amount of nitrogen in the blood from urea, a waste product of protein breakdown. BUN is used to assess kidney function, diagnose kidney disease or dehydration, and monitor patients with kidney disease or who are on dialysis.

Figures 3-17 and **3-18** list the causes of an increased or decreased BUN.

FIGURE 3-17 CAUSES OF INCREASED BUN

INCREASED PROTEIN METABOLISM

Increased protein intake

Increased protein breakdown

- *Upper GI hemorrhage*: Increased protein breakdown + decreased blood volume
- *Infection*
- *Fever*
- *Excess exercise*
- *Shock*
- *Stress and surgery*
- *Burns*
- *Starvation*
- *Hyperthyroidism* (increased metabolic rate leads to increased protein breakdown)

PRERENAL FAILURE (less renal perfusion results in less urea being filtered, increasing BUN)

Volume depletion

- *Dehydration*

Effective volume depletion (the total body fluid volume may be normal but is not effectively circulating in the arteries)

- *Congestive heart failure:* The heart doesn't pump effectively, leading to reduced cardiac output and poor tissue perfusion.
- *Cirrhosis:* Fluid is displaced into ascites.

INTRINSIC RENAL FAILURE (less renal perfusion results in less urea being filtered, increasing BUN)

- *Acute tubular necrosis*
- *Drugs*
- *Ischemia*
- *Toxins (amyloid, myoglobin)*
- *Infection (pyelonephritis)*
- *Nephrotic syndrome with volume depletion:* Over time, increased glomerular permeability causes decreased glomerular filtration rate (GFR) with backup of BUN, along with the proteinuria, hypoalbuminemia and edema.
- *Nephritic syndromes:* Glomerular damage and inflammation cause decreased glomerular filtration rate, with hematuria, hypertension, and proteinuria.

POSTRENAL FAILURE: Obstruction results in back pressure that reduces the ability of kidney to filter BUN.

- *Nephrolithiasis*
- *Tumor*
- *Congenital anomaly*
- *Post surgical* (e.g., ureteral stricture)

FIGURE 3-18 CAUSES OF DECREASED BUN

DECREASED PROTEIN INTAKE

- *Malnutrition*
- *Malabsorption*

INCREASED PROTEIN LOSS

- *Dialysis*: May lower BUN by removing urea from the blood.
- Protein-losing enteropathy

DECREASED BUN PRODUCTION

- *Liver disease*: The liver normally produces BUN from urea. BUN may thus be decreased in liver disease.

DILUTION

- *Overhydration*
- *Syndrome of Inappropriate Antidiuretic Hormone secretion* (SIADH): Dilutes the BUN
- *Pregnancy*: BUN dilution through increased plasma volume; increased GFR clears more BUN; increased delivery of amino acids to the fetus rather than for breakdown to BUN

Creatinine

Creatinine levels help assess kidney function and the effectiveness of treatment for kidney disease. Creatinine is a nitrogenous waste product of the breakdown of muscle creatine. BUN is a waste product of protein breakdown. While both BUN and creatinine are useful in assessing renal function, BUN is more useful in detecting early renal changes, but is also influenced by changes in protein intake, catabolic states, and liver function. Creatinine is more specific for renal function but is affected by muscle mass and is a later indicator of kidney dysfunction (≥ 50% loss of function) than is BUN.

The **BUN/creatinine** ratio (normally about 10:1 to 20:1) typically increases in conditions where there is increased urea production (increased protein intake or breakdown to BUN) or decreased renal perfusion. Why does renal perfusion affect the BUN/creatinine ratio? Urea normally is much better reabsorbed in the kidney than creatinine. In prerenal failure, where there is decreased renal perfusion, the preference for urea reabsorption over that of creatinine results in a higher BUN/creatinine ratio (often > 20:1).

The **estimated Glomerular Filtration Rate (eGFR)** is a calculated value that considers serum creatinine levels as well as age, sex, and sometimes race. It gives an estimation of how well the kidneys are filtering waste products from the blood and provides an index of kidney health. Poor filtration could be due to intrinsic renal, prerenal or postrenal failure:

Intrinsic renal failure results from direct injury to the kidneys, with reduced glomerular filtration rate most often due to ischemia, toxins, inflammation, or immune-mediated processes affecting the tubules, interstitium, glomerulae, or vessels.

Prerenal failure is marked by reduced glomerular filtration rate that may be due to volume loss, heart failure, hypotension, renal stenosis, or drugs affecting renal autoregulation. The kidneys themselves may be normal, but blood is not reaching the kidneys well.

Postrenal failure results from reduced glomerular filtration rate caused by urinary backflow from urinary tract stones, tumors, prostate enlargement, catheter obstruction, or strictures. Again, the kidneys themselves may be normal.

Figures 3-19 and **3-20** list the causes of an increased or decreased creatinine.

FIGURE 3-19 CAUSES OF INCREASED CREATININE
INCREASED CREATININE PRODUCTION • *Increased muscle mass or activity* • *Muscle injury* • *Medications that may cause rhabdomyolysis (rapid muscle breakdown):* E.g., statins • *Creatine supplements:* Used to increase muscle mass. Creatinine is a breakdown product of creatine. **PRERENAL FAILURE** ***Volume depletion*** • *Dehydration* • *emorrhage* ***Effective volume depletion*** (volume, while not depleted, is not detected by the kidney) • Congestive heart failure • Cirrhosis with ascitic pooling • Septic shock **INTRINSIC RENAL FAILURE** • *Acute tubular necrosis* (from drugs, ischemia, or toxins) • *Infection* (pyelonephritis) • *Nephrotic syndromes* • *Nephritic syndromes* **POSTRENAL FAILURE** ***Obstruction*** • *Nephrolithiasis* • *Tumor* • *Congenital anomaly* • *Post surgical* (e.g., ureteral stricture)

FIGURE 3-20 CAUSES OF DECREASED CREATININE
DECREASED MUSCLE MASS: Creatinine is a waste product of creatine catabolism in muscle contraction. • *Muscle-wasting diseases* • *Muscular dystrophy* • *Myasthenia gravis* • *Small stature* • *Debilitation* • *Aging* • *Prolonged bed rest* **DILUTION** • *Overhydration* • *Pregnancy* (dilution from increased plasma volume; increased glomerular filtration rate with loss of creatinine)

Creatinine should not be confused with the enzyme **creatine phosphokinase (CPK; CK)**:

- **Creatinine** is a normal waste product of muscle creatine energy metabolism that is excreted by the kidney; blood creatinine levels are often used to assess kidney function when the blood levels of creatinine back up and rise.
- **CPK**, though, is an important enzyme found (with variations), in the heart (CK-MB), brain (CK-BB), and skeletal muscles (CK-MM); CPK is important in the conversion of creatine for energy production. Elevated blood levels of CPK may indicate muscle damage or cardiac infarction when the CPK leaks out of damaged cells into the bloodstream. CPK levels may be elevated after vigorous exercise (indicating muscle damage).

Although elevated CK-MB is a useful indicator of myocardial infarction, elevated cardiac **troponins** (used in muscle contraction) are a more specific and sensitive indicator of cardiac injury when they leak out of cardiac muscle cells.

BNP (B-type Natriuretic Peptide) is a commonly used marker of cardiac dysfunction. Unlike CK-MB, which rises due to leakage from damaged cardiac cells, BNP is actively *secreted by the ventricles* in response to increased pressure and volume overload, which stretch the heart muscle. BNP promotes *vasodilation* and *renal sodium excretion*, helping to reduce blood pressure and intravascular volume.

Elevated levels of **BNP** or **NT-proBNP** can help distinguish *heart failure* from *pulmonary causes* of dyspnea. BNP levels correlate roughly with the severity of heart failure and have modest prognostic significance. Levels typically decrease with effective heart failure treatment and may be useful in monitoring disease progression. Notably, BNP levels may be *falsely low in obese individuals*.

Liver Function Tests

Liver function tests are used to evaluate liver/biliary system health and function: **Total protein**, **albumin**, and **globulin** are important products produced by the liver. **ALT**, **AST**, and **ALP** are enzymes produced by the liver/biliary system that leak out of cells when the cells are injured and serve as markers for diagnosing liver or biliary injury. **Bilirubin** results from the normal breakdown of red blood cells. Although it is primarily a waste product, it also acts as an antioxidant throughout the body, where it helps protect cells from oxidative damage.

Bilirubin

There are two forms of bilirubin: **unconjugated** (**indirect**) and **conjugated** (**direct**). The unconjugated form results from the breakdown of heme in red blood cells during RBC destruction. Unconjugated bilirubin is not water-soluble or excretable in the urine; it circulates in the bloodstream bound to albumin, where it is carried to the liver to be processed into conjugated (direct) bilirubin. Conjugated bilirubin is secreted into the bile ducts; it is water soluble, and excess conjugated bilirubin can be excreted in the urine. An excess of either type of bilirubin can cause jaundice.

Figure 3-21 shows points of interruption in bilirubin processing where diseases can occur.

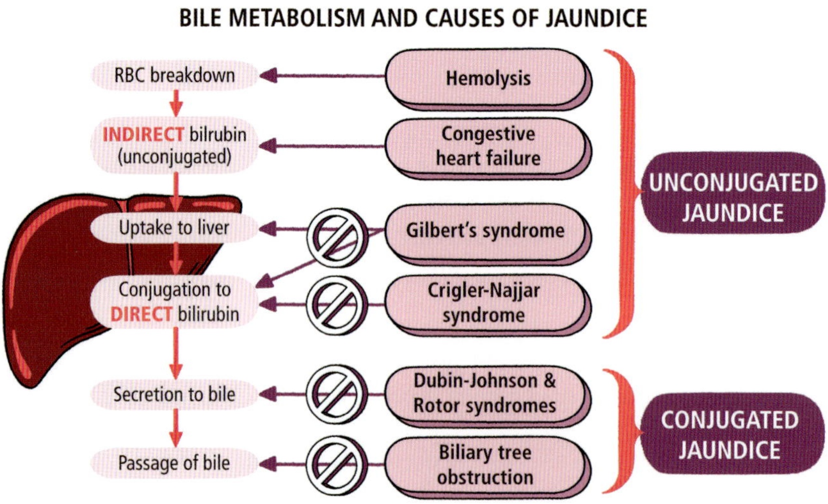

Figure 3-21 *Points of pathology in bile metabolism. (Modified from Berkowitz, A. Clinical Pathophysiology Made Ridiculously Simple, Medmaster)*

Figure 3-22 lists the potential causes of hyperbilirubinemia.

FIGURE 3-22 CAUSES OF HYPERBILIRUBINEMIA
INCREASE IN UNCONJUGATED (INDIRECT) BILIRUBIN ***Increased production of unconjugated bilirubin***: Hemolysis causes an increase in unconjugated bilirubin (not excreted in urine). • *Hemolytic anemia* • *Erythroblastosis fetalis* • *Transfusion hemolysis*

- *Sickle cell anemia*
- *Pernicious anemia:* The large, immature RBCs are fragile and subject to hemolysis.
- *Hematoma breakdown*

Decreased delivery of bilirubin to the liver
- *CHF*
- *Surgical shunts*

Decreased liver uptake of bilirubin
- *Gilbert's syndrome*

Decreased conjugation of bilirubin
- *Gilbert's syndrome*
- *Crigler-Najjar syndrome*
- *Neonatal hyperbilirubinemia* (liver not mature enough to handle the load of bilirubin conjugation)

INCREASE IN CONJUGATED (DIRECT) BILIRUBIN

Liver disease (decreases liver's ability to secrete the conjugated bilirubin)
- *Hepatitis:* The damaged liver cells can still conjugate bilirubin but can't excrete it into the bile ducts.
- *Cirrhosis:* Scarring blocks the excretion of bilirubin-containing bile into the bile ducts.

Failure to secrete conjugated bilirubin from the liver
- *Dubin-Johnson syndrome*: Poor secretion of conjugated bilirubin from hepatic cells
- *Rotor syndrome*: Poor secretion of conjugated bilirubin from hepatic cells

Biliary obstruction
- *Liver metastases*
- *Drug cholestasis side effect:* Bile cannot be excreted from the hepatocytes.
- *Inflammation of biliary ducts (primary biliary cirrhosis; primary sclerosing cholangitis)*
- *Gallstones*
- *Tumors*
- *Surgical scarring*

Alanine Aminotransferase (ALT)

ALT levels in the blood mainly provide information about the liver, which has the highest concentration of ALT, even though ALT is widely distributed in small amounts in the body, e.g., muscles, heart, kidneys, pancreas, and RBCs. ALT is an enzyme that is detected when the cells are injured and the ALT leaks out.

Figure 3-23 lists the causes of increased ALT.

FIGURE 3-23 CAUSES OF INCREASED ALT (SGPT)

LIVER DISEASE: ALT is released from damaged liver cells.

- *Hepatitis*
- *Cirrhosis*
- *Liver tumors*
- *Fatty liver disease*
- *Drug toxic effect on liver*
- *Alcoholism*

TRAUMA TO HEART

- *Myocardial infarction:* Causes a rise in troponins, CK-BM, and LDH, which leak out of damaged cardiac cells. A rise in ALT may occur as a secondary effect on the liver (e.g., hypoxia)
- *Heart surgery:* When the liver is affected (e.g., reduced perfusion)

INJURY TO SKELETAL MUSCLE: ALT is released from damaged muscle cells.

- *Trauma to muscle*
- *Muscular dystrophy*
- *Myopathies*
- *Myositis*

OTHER:

- *Hemolytic anemia:* ALT is increased due to increased workload on the liver, which has to process bilirubin from the hemolysis.
- *Megaloblastic anemia:* The liver is stressed from processing the excess breakdown products from the turnover and destruction of immature red blood cells in the bone marrow.
- *Malignancy*
- *Acute pancreatitis:* Can result in liver damage
- *Lung infarct:* Hypoxia or infection can affect the liver.
- *Kidney disease:* Accumulation of toxins can tax the liver.
- *Burns:* Inflammation and shock can affect the liver.
- *Seizures:* Muscle damage with release of ALT
- *Eclampsia:* Liver damage from severe hypertension and microthrombi in liver blood vessels with ischemia and hepatocellular injury
- *Severe infections:* Systemic inflammation can affect the liver.
- *Drugs:* E.g., statins, antibiotics, anticonvulsants, and chemotherapeutic agents can be hepatotoxic.

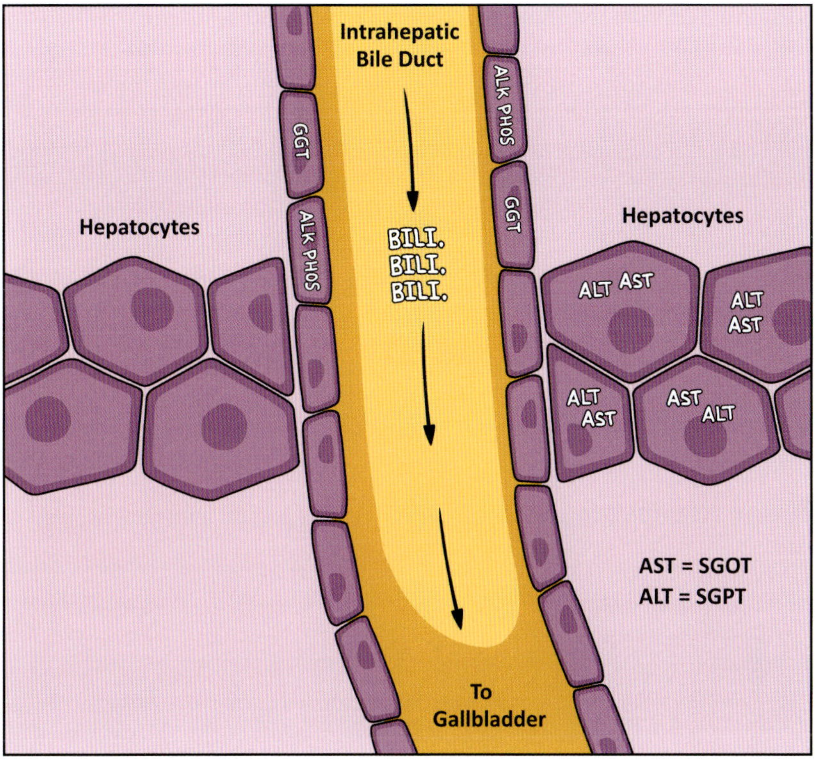

Fig. 3-24 *Location of ALT, AST, alkaline phosphatase (ALP), and gamma-glutamyl transferase (GGT) in the hepatobiliary system. The bile ducts have a relatively greater amount of ALP and GGT. A prominent rise in ALP and GGT commonly suggests biliary tract obstruction.*

Aspartate Aminotransferase (AST)

Like alanine aminotransferase (ALT), AST can be found in a variety of tissues, including the liver, heart, and skeletal muscle and can be elevated when these tissues are injured. AST, though, is found in higher concentrations than ALT in other tissues than the liver. ALT is more specific to the liver. Elevation in AST with relatively normal ALT can suggest involvement of other organs besides the liver.

Figure 3-25 lists the main causes of an increased AST.

FIGURE 3-25 CAUSES OF INCREASED AST (SGOT)
LIVER DISEASE: AST is released from damaged liver cells. • *Hepatitis* • *Cirrhosis* • *Liver cancer* • *Fatty liver disease* • *Drug toxic effect on liver* • *Alcoholism*
TRAUMA TO HEART: AST is released from damaged myocardial cells. • *Myocardial infarction* • *Cardiac catheterization* • *Heart surgery*
INJURY TO SKELETAL MUSCLE: AST is released from damaged muscle cells. • *Trauma to muscle* • *Muscular dystrophy* • *Myopathies* • *Myositis*
OTHER: AST is widely distributed in the tissues and may leak out during injury. • Acute pancreatitis • *Lung infarct* • *Renal infarct* • *Burns* • *Seizures* • *Brain injuries* • *Eclampsia* • *Severe infections* • *Drugs:* E.g., statins, acetaminophen, NSAIDs, some antibiotics, anticonvulsants, chemotherapeutic agents, anabolic steroids, some herbal supplements

Alkaline Phosphatase (ALP)

While elevation of alanine aminotransferase (ALT) and aspartate aminotransferase (AST) mainly indicate liver damage, alkaline phosphatase (ALP) is also found in high concentration in bile ducts (**Fig. 3-24**) and bone. Elevated ALP often indicates obstruction in bile ducts or bone damage, which may not present with elevated ALT or AST. When ALP is elevated, **gamma-glutamyl transferase (GGT)** is often measured to confirm hepatobiliary obstruction, since GGT is also prominent in biliary ducts (**Fig. 3-24**). A normal GGT along with an elevated ALP suggests the problem may lie in the bone.

Figures 3-26 and **3-27** list the causes of an increased or decreased alkaline phosphatase.

FIGURE 3-26 CAUSES OF INCREASED ALKALINE PHOSPHATASE (ALP)

BILIARY DUCT OBSTRUCTION and to a lesser degree, hepatic disease

Intra or extrahepatic biliary duct obstruction

- *Gallstone*
- *Cholecystitis*
- *Primary sclerosing cholangitis*
- *Tumors*

Liver diseases

- *Cirrhosis*
- *Diabetic hepatic lipidosis*
- *Sarcoid*
- *Tuberculosis*
- *Amyloidosis*
- *Liver abscess*
- *Drug-induced liver disease*
- *Hepatitis*
- *Liver tumors*

ALKALINE PHOSPHATASE FROM BONES: From osteoblasts

- *Metastatic disease to bone*
- *Osteomalacia/rickets*
- *Paget's disease of bone*
- *Osteogenic sarcoma*
- *Leukemia*
- *Myelofibrosis*: Fibrous tissue replaces the blood-producing cells in the bone marrow.
- *Fractures*
- *Hyperparathyroidism*
- *Hyperthyroidism:* Promotes bone turnover, with ALP coming from osteoblasts
- *Rheumatoid arthritis*
- *Bone growth spurt in children*
- *Acromegaly*
- *Pregnancy, 3rd trimester*: The placenta produces ALP; bone growth of fetus
- *Hypervitaminosis D*

OTHER

- *Familial hyperphosphatasia:* Causes increased production or decreased degradation of ALP

FIGURE 3-27 CAUSES OF DECREASED ALKALINE PHOSPHATASE
HYPOPHOSPHATEMIA: Phosphate is necessary to make alkaline phosphatase. • *Poor nutrition* • *Increased urine excretion of phosphate* • *Decreased intestinal absorption of phosphate* • *Redistribution of phosphate* (respiratory alkalosis can cause a shift of phosphate into cells) **OTHER**: • *Hypothyroidism* (thyroid hormones play a role in bone turnover and liver function) • *Pernicious anemia* (B12 deficiency can affect ALP activity) • *Vitamin C deficiency* (impaired collagen and bone formation) • *Adrenal insufficiency* (lack of adrenal hormones can affect liver and bone metabolism) • *Zinc deficiency* (zinc is an important cofactor in ALP synthesis)

Total Protein

Albumin in the most common protein in the blood. Other important blood proteins include immunoglobulins, acute phase proteins (e.g., C-reactive protein, amyloid A, fibrinogen), transport proteins (e.g., haptoglobin, ceruloplasmin, transferrin), clotting factors, complement proteins (involved in the immune response) and others. The total protein test typically measures the levels of albumin and globulins, which are the main protein components in the blood.

Many serum proteins increase in response to inflammation or infection, or the need for clotting factors or transport.

Figures 3-28 and **3-29** list the causes of hyperproteinemia and hypoproteinemia. (**Figures 3-30** and **3-31** specifically refer to albumin, **Figs. 3-32** and **3-33** to globulins.

FIGURE 3-28. CAUSES OF HYPERPROTEINEMIA
HYPERGAMMAGLOBULINEMIAS ***Monoclonal hypergammaglobulinemias:*** Overproduction of a single type of immunoglobulin by a single clone of plasma cells • *Malignant hypergammaglobulinemias:* Arise from malignant proliferation of a single type of plasma cells or B cell (e.g., multiple myeloma; plasmacytoma; amyloidosis; Waldenstrom's macroglobulinemia; malignant lymphoma; heavy chain disease; monoclonal gammopathy of unknown significance)

Polyclonal hypergammaglobulinemias: Overproduction of multiple types of immunoglobulins by various plasma cell clones

- *Benign hypergammaglobulinemias:* Immunoglobulin rise is not associated with malignancy (e.g., infections, autoimmune diseases)
- *Collagen diseases* (systemic lupus erythematosus; rheumatoid arthritis; scleroderma)
- *Liver disease* (cirrhosis and chronic hepatitis can cause overproduction of certain proteins despite the liver dysfunction)
- *Chronic infections* (tuberculosis; bronchiectasis; subacute bacterial endocarditis; lung abscess; osteomyelitis)
- *Chronic inflammatory diseases*: E.g., inflammatory bowel disease
- *Sarcoidosis:* Stimulates protein production through chronic inflammation and activation of the immune system
- *Certain medications:* E.g., certain antibiotics, anticonvulsants, antiviral drugs, immunosuppressants

ACUTE PHASE PROTEINS

- *C-reactive protein and serum amyloid A:* Acute-phase proteins that are markers of inflammation
- *Fibrinogen:* An essential plasma clotting factor that converts to fibrin during coagulation

TRANSPORT PROTEINS

- *Haptoglobin:* Binds free hemoglobin from lysed RBCs and brings it to the liver to form bilirubin
- *Ceruloplasmin:* A copper-carrying protein
- *Transferrin:* Transports iron in the bloodstream

CLOTTING FACTORS: E.g., fibrinogen, prothrombin, thromboplastin, and others

COMPLEMENT PROTEINS: E.g., C3 and C4

OTHERS: E.g., *alpha- antitrypsin*, (protects the lungs from enzyme damage); *beta-2 microglobulin* (part of the immune system)

DEHYDRATION: Concentrates the protein, including albumin, although the actual amount of protein may remain the same.

FIGURE 3-29. CAUSES OF HYPOPROTEINEMIA
POOR PROTEIN INTAKE • *Malnutrition* • *Malabsorption*
EXCESS PROTEIN LOSS • *Nephrotic syndrome* • *Burns* • *GI disorders* (protein-losing enteropathies; surgical resection) • *Blood loss* • *Plasmapheresis:* Plasma removal removes proteins. • *Pregnancy:* Loss in urine; also, increased protein requirements and use by the fetus and mother • *Ascites*: Albumin moves from blood to peritoneal space. • *Lupus erythematosus*: If there is loss of albumin in urine • *Hyperthyroidism:* Increased metabolic rate causes increased protein breakdown. • *Chronic infection and inflammation:* Can increase protein breakdown • *Hemorrhage*
DEFECTIVE PROTEIN SYNTHESIS • *Liver disease*: Decreased albumin synthesis • *Agammaglobulinemia* • *Familial idiopathic dysproteinemia*
DILUTION EFFECT • *Excess IV administration* • *Syndrome of inappropriate ADH secretion*

Albumin

Albumin is the most abundant protein in the serum. It is important in transporting many substances in the blood, in addition to maintaining plasma oncotic pressure.

Figures 3-30 and **3-31** list the causes of hyperalbuminemia and hypoalbuminemia.

FIGURE 3-30 CAUSES OF HYPERALBUMINEMIA
DEHYDRATION: Albumin is then concentrated.
EXCESS INTAKE OF ALBUMIN: Albumin infusion

FIGURE 3-31 CAUSES OF HYPOALBUMINEMIA
POOR PROTEIN INTAKE ***Malnutrition*** ***Malabsorption*** • *Crohn's disease* • *Sprue* • *Whipple's disease*: An infectious disease of the small bowel that causes malabsorption **POOR PROTEIN SYNTHESIS** (albumin is produced in the liver) ***Liver disease*** • *Hepatitis* • *Cirrhosis* • *Metastatic tumor to liver* • *Chronic inflammation:* Can decrease albumin synthesis and increase albumin breakdown, prioritizing the synthesis of acute-phase proteins • *Hypothyroidism:* Metabolic slowdown affects albumin synthesis. **INCREASED PROTEIN LOSS** ***Protein loss through kidneys*** • *Nephrosis:* A non-inflammatory kidney disease, such as nephrotic syndrome, where protein is lost in the urine • *Pregnancy:* Increased kidney filtration of protein and increased use of protein for fetal growth ***Protein loss through skin lesions*** ***Ascites***: Albumin moves from blood to peritoneal space. ***Protein-losing enteropathies*** ***Increased protein catabolism*** • *Burns* • *Familial idiopathic dysproteinemia*: A hypercatabolic hypoproteinemia **DILUTIONAL** • *Excess hydration* • *Heart failure*: Fluid retention and dilution of albumin

Globulin

Globulins are proteins that are important in the immune system (immunoglobulins), blood clotting (fibrinogen), and in maintaining oncotic pressure along with albumin. They also act as transport proteins (e.g., for hormones, metals, lipids, vitamins, hemoglobin, and immune complexes). Globulin is often calculated by subtracting albumin from the total protein.

Figures 3-32 and **3-33** list the causes of hyperglobulinemia and hypoglobulinemia.

FIGURE 3-32 CAUSES OF HYPERGLOBULINEMIA

ALPHA-GLOBULIN INCREASED: Alpha-globulins are acute phase reactant proteins that help modulate the immune response.

Alpha1-globulin: E.g., alpha1-antitrypsin, which protects tissue from the enzymes of inflammatory cells, especially elastase

- *Inflammation*
- *Infarction*
- *Cancer*
- *Burns*
- *Necrosis*
- *Pregnancy:* Acute phase proteins help protect the mother and fetus

Alpha2-globulin: An acute phase reactant and transport protein. Haptoglobin is a globulin that binds to hemoglobin, preventing kidney damage, and brings hemoglobin to the liver and spleen for clearance. Ceruloplasmin is a copper-binding globulin.

- *Inflammation*
- *Nephrotic syndrome*: Alpha2-globulin may increase to compensate for loss of protein in the urine,
- *Adrenal insufficiency*: Alpha2-globulin synthesis is an attempt to compensate for the deficiency of corticosteroids.
- *Corticosteroid therapy*: Induces the liver to produce more acute-phase proteins, like alpha2-globulin
- *Advanced diabetes mellitus*: Chronic inflammation can stimulate the liver to produce more acute-phase proteins.

BETA-GLOBULIN INCREASED: Some beta-globulins (beta-lipoproteins) are involved in the transport of lipids and cholesterol; others in iron transport (transferrin); blood clotting (plasminogen, Factor IX); and the immune response (many complement proteins).

- *Primary hypercholesterolemia:* Beta-globulins are needed to transport cholesterol in the blood.
- *Hypothyroidism:* Reduced lipid metabolism results in increased LDL and more beta-globulins to carry the LDL.
- *Biliary cirrhosis:* Decreased ability to clear certain proteins, including globulins, from the blood, leading to their accumulation
- *Nephrotic syndrome:* Loss of protein in the urine leads to a compensatory increase of certain proteins, including beta-globulin.
- *Malignant hypertension:* Inflammatory damage to blood vessel stimulates synthesis of acute-phase proteins like beta-globulins.

- *Cushing's disease:* Increased cortisol increases LDL levels.
- *Polyarteritis nodosa:* Inflammation leads to increase synthesis of acute-phase reactants.
- *Pregnancy - third trimester:* Pregnancy is accompanied by increased synthesis of various proteins, including beta-globulins.
- *Iron-deficiency anemia:* Transferrin is increased.

GAMMA-GLOBULIN INCREASED: (All gamma-globulins are immunoglobulins)

Chronic inflammation: Autoantibodies are gamma-globulins
- *Rheumatoid arthritis*
- *Lupus erythematosus*

Acute and chronic infections

Multiple myeloma: Excess plasma cells and gamma-globulin produced by them

Certain cancers: Overproduction of immunoglobulins by malignant or abnormal plasma cells
- *Chronic lymphocytic leukemia*
- *Malignant lymphoma*
- *Lymphoma*
- *Hodgkin's disease*
- *Waldenstrom's macroglobulinemia*

Cirrhosis: Impaired ability to clear immunoglobulins. Also, chronic inflammation from liver damage spurs the production of immunoglobulins.

Autoimmune hepatitis

Amyloidosis: Chronic inflammation

FIGURE 3-33 CAUSES OF HYPOGLOBULINEMIA

ALPHA-GLOBULIN DECREASED: Alpha-globulins are acute phase reactant proteins that help modulate the immune response.

Alpha1-globulin: E.g., alpha1-antitrypsin, which protects tissue from the enzymes of inflammatory cells, especially elastase
- *Juvenile pulmonary emphysema*: Alpha1-antitrypsin deficiency

Alpha2-globulin: An acute phase reactant and transport protein. Haptoglobin is a globulin that binds to hemoglobin, preventing kidney damage and helping the removal of hemoglobin by the liver. Ceruloplasmin is a copper-binding globulin.
- *Wilson's disease*: Ceruloplasmin levels are low in Wilson's disease and there is subsequent abnormal copper accumulation in the liver, brain, eyes, kidneys, and joints.

- *Hyperthyroidism:* Causes increased catabolism of proteins, including globulins.
- *Hemolysis*: Haptoglobin binds to free hemoglobin and is rapidly cleared, lowering haptoglobin.
- *Severe liver disease:* Impairs the ability of the liver to synthesize proteins, including globulins
- *Protein-losing enteropathies:* Protein loss includes globulins.
- *Malnutrition:* Inadequate protein intake
- *Megaloblastic anemia:* B12 deficiency affects protein synthesis in different cells, including those that produce plasma proteins.

BETA-GLOBULIN DECREASED: Beta globulins are involved in lipoprotein transport (lipoproteins) and iron transport (transferrin); they are also part of plasminogen and complement

- Malnutrition

GAMMA-GLOBULIN DECREASED: All gamma globulins are immunoglobulins.

Various genetic immune disorders

- *Agammaglobulinemia*

Secondary immune deficiency

- *Leukemia*
- *Steroid overuse*
- *Severe infection:* Immune system exhaustion

Nephrotic syndrome: Protein lost in the urine

The Albumin/Globulin (A/G) Ratio

The A/G ratio, usually slightly above 1, provides a useful assessment of liver and renal function, as well as infections and immune disorders. The liver is the main producer of albumin. Although the liver also produces some globulins, most globulins are produced by the immune system. Hence:

- A *decreased* A/G ratio occurs in liver disease, since the liver produces less albumin. The A/G is also decreased in chronic infections, inflammation, or autoimmune disease due to the increased synthesis of globulins (immunoglobulins). The A/G may also be decreased in renal disease where there is loss of protein, particularly albumin, the most abundant protein in the blood.
- An *increased* A/G ratio occurs in immunodeficiency due to decreased production of immunoglobulins.

Hepatitis Virus Tests

Hepatitis is an inflammation of the liver, often caused by viral infections, but also by other things such as excess alcohol use, toxins, medications, and certain medical conditions, such as hemochromatosis, Wilson's disease, and alpha-antitrypsin deficiency.

In a nutshell:

- There are different types of viral hepatitis, including hepatitis A, B, C, D, and E, each caused by a different virus.
- The "ends," A and E, of the letter sequence "ABCDE," correspond to hepatitis A and E being spread by the oral/fecal route, the "ends" of the GI tract. The others, BCD, are spread by blood-to-blood contact (e.g., needles) or exchange of bodily fluids (e.g., unprotected sexual contact).
- While A, B, C, D, and E can all cause acute hepatitis. B, C, and D can also cause *chronic* hepatitis and are risk factors for the development of cirrhosis and hepatocellular carcinoma. (Hepatitis D can only infect people who are also infected with hepatitis B.)
- Hepatitis C is the leading cause of chronic hepatitis in the U.S. and the leading cause of liver transplantation. Fortunately, hepatitis C can be cured with newer antivirals.
- Hepatitis B can be treated with interferons and nucleoside analogs, but there is still a possibility of relapse.
- There are vaccines effective for preventing hepatitis A and B.
- In general: The presence of viral antigens, viral antibodies, and viral genetic material in the blood quantified by PCR (polymerase chain-reaction) helps determine whether an individual has been exposed to virus, immunized against virus, or whether there is dormant or active infections. The presence of viral antigens or viral genetic material in the blood indicates active infection and potential infectivity. The patient who has viral antigens for hepatitis A or B, or antibodies in hepatitis C is considered potentially infectious. The presence of antibodies only (no viral antigens or viral genetic material) indicates past exposure or immunity and lack of infectivity in hepatitis A or B. In the case of hepatitis C (HCV) the antigen test is not very sensitive; the patient may still be infectious with HCV even if HCV antigen is not detected. The HCV RNA test by PCR (polymerase chain reaction) can detect lower levels of the virus than the antigen test. A test positive for anti-HCV antibodies and negative for HCV RNA indicates that the patient has been exposed to HCV but is not infectious. If both are positive, it indicates active infection, even if the antigen test for HCV is negative. (See **Figures 3-34A**, **3-34B**, and **3-34C** for more details).

FIGURE 3-34A HEPATITIS A VIRUS TESTING		
Test	**Result**	**Interpretation**
Anti-HAV IgM	Positive	Acute or recent Hepatitis A infection.
Anti-HAV IgM	Negative	No acute Hepatitis A infection.
Anti-HAV Total (IgG)	Positive	Past infection or immunity due to vaccination.
Anti-HAV Total (IgG)	Negative	No prior exposure or immunity to Hepatitis A.

FIGURE 3-34B HEPATITIS B VIRUS TESTING		
Test	**Result**	**Interpretation**
HBsAg (Surface Antigen)	Positive	Active infection (acute or chronic).
HBsAg	Negative	No active Hepatitis B infection.
Anti-HBs (Surface Antibody)	Positive	Immunity due to vaccination or recovery from past infection.
Anti-HBs	Negative	No immunity to Hepatitis B.
Anti-HBc (Core Antibody)	Positive	Past or ongoing Hepatitis B infection.
Anti-HBc IgM	Positive	Acute Hepatitis B infection.
HBV DNA (Viral Load)	Detectable	Active viral replication.
HBV DNA (Viral Load)	Undetectable	No active viral replication.

FIGURE 3-34C HEPATITIS C VIRUS TESTING		
Test	**Result**	**Interpretation**
Anti-HCV (Antibody)	Positive	Current or past Hepatitis C infection.
Anti-HCV	Negative	No evidence of Hepatitis C exposure.
HCV RNA (Viral Load)	Detectable	Active Hepatitis C infection (confirmatory test).
HCV RNA (Viral Load)	Undetectable	No active Hepatitis C infection.
HCV Genotype	Genotype 1-6	Determines the strain of HCV for treatment planning.

Vitamins

Although a vitamin screening panel (for vitamins B12, folate, D, E, A, K, and C) is not routine, perhaps it should be. This is not just to check for vitamin deficiency, but for vitamin overdose; vitamin supplements today are commonly taken in excess, which can cause clinical problems that sometimes have nonspecific symptoms and signs (**Fig. 3-3**).

FIGURE 3-35 SIGNS OF VITAMIN DEFICIENCY AND EXCESS		
Vitamin	**Signs of Deficiency**	**Signs of Excess**
B12	Fatigue, weakness, decreased appetite, weight loss, constipation, megaloblastic anemia, mouth sores, neurologic problems with balance, depression, and memory	Generally safe, but high doses over prolonged periods can cause dizziness, headache, anxiety, nausea, or skin rash
Folate	Symptoms like those in B12 deficiency, including neural tube defects in the fetus	Can mask vitamin B12 deficiency; bloating, gas, nausea
D	Bone pain, muscle weakness, osteoporosis, fractures. Rickets in children, osteomalacia in adults	Hypercalcemia, kidney damage, nausea, weakness, cardiovascular problems
E	Muscle weakness, balance difficulty, numbness, retinopathy	Increased risk of hemorrhage; nausea, diarrhea, fatigue, weakness, headache, blurred vision, rashes
A	Dry eyes (xerophthalmia) and dry skin, night blindness, delayed growth	Nausea, headache, fatigue, decreased appetite, dizziness, dry skin; liver damage; teratogenic effects in pregnant women
K	Impaired clotting, excess bleeding	Jaundice, hemolytic anemia in infants; interference with anticoagulant medications in adults
C	Scurvy (anemia, fatigue, bleeding, pain in the limbs, ulceration of the gums, and tooth loss)	Diarrhea, nausea, abdominal cramps, kidney stones (oxalate crystals)

Commonly, a vitamin *excess* results from excessive supplement intake, while a vitamin *deficiency* results from poor dietary intake or malabsorption. For instance, vitamins D, E, A, and K are fat-soluble, and may be deficient with conditions of fat malabsorption. Certain medications can cause deficiency in a vitamin by decreasing its absorption, increasing its excretion, or otherwise altering its metabolism.

4

Lipid Panel

The lipid panel is used to assess cardiovascular health, which may be affected by blood lipid levels. The panel typically includes total cholesterol, low-density lipoprotein (LDL) cholesterol, high-density lipoprotein (HDL) cholesterol, and triglycerides. Increased levels of LDL and triglycerides, along with low HDL, are associated with an increased risk of atherosclerosis, heart attacks, and strokes.

Total Cholesterol

Cholesterol (the sum of HDL, LDL, and VLDL cholesterol) is an important part of cell membranes and is a precursor to steroid hormones, among other things. Hypercholesterolemia (particularly increased LDL) increases the risk of atherosclerosis and cardiovascular disease.

Figures 4-1 and **4-2** list the causes of hypercholesterolemia and hypocholesterolemia.

FIGURE 4-1 CAUSES OF HYPERCHOLESTEROLEMIA

EXCESS DIETARY INTAKE OF SATURATED FAT/CHOLESTEROL

- *High intake of saturated fats, trans fats (unsaturated fatty acids with at least one double bond in the trans configuration), and cholesterol-rich foods*

EXCESS SYNTHESIS OF CHOLESTEROL

- *Familial hypercholesterolemia* (especially LDL)
- *Polygenic hypercholesterolemia:* High cholesterol due to multiple genetic mutations in an individual
- *Familial hyperlipidemia:* Elevated levels of various lipids, including cholesterol and triglycerides
- *Werner's syndrome:* A genetic disease of accelerated aging, with increased levels of cholesterol and other lipids

DECREASED REMOVAL OF CHOLESTEROL

- *Biliary obstruction*
- *Anorexia nervosa*

SECONDARY FACTORS:

- *Obesity:* Associated with high cholesterol; more fatty acids available to form triglycerides and cholesterol
- *Lack of physical exercise:* Weight gain
- *Smoking:* Can increase LDL and decrease HDL
- *Hypothyroidism:* Reduced metabolism of lipids
- *Diabetes mellitus:* Loss of insulin activity results in increased free fatty acid release to produce energy.
- *Polycystic ovary syndrome:* Associated with insulin resistance
- *Pregnancy:* Estrogen stimulates the liver to produce more cholesterol for fetal development and placental function.
- *Nephrotic syndrome:* Loss of protein causes the liver to compensate by increasing synthesis of proteins, including lipoproteins, which carry cholesterol and triglycerides in the blood; there may also be reduced clearance of lipoproteins by the kidney.
- *Age and sex:* Cholesterol tends to increase with age; men typically have higher cholesterol levels than women until menopause, as estrogen has a protective effect on lipid metabolism.
- *Medications:* Corticosteroids, some diuretics, beta-blockers, and excess alcohol can result in a higher cholesterol.

FIGURE 4-2 CAUSES OF HYPOCHOLESTEROLEMIA
DECREASED INTAKE OF FAT/CHOLESTEROL: Cholesterol is largely synthesized from fat obtained in the diet. • *Decreased dietary intake of fats* • *Malabsorption:* E.g., celiac disease, Crohn's disease can impair the absorption of fats and cholesterol; abetalipoproteinemia is an inherited disorder that affects fat/cholesterol absorption. **DEFECT IN CHOLESTEROL SYNTHESIS** • *Liver disease*: Impairment of the liver to synthesize cholesterol • *Tangier disease:* There is low HDL from impaired transport of cholesterol out of cells. • *Hyperthyroidism*: Increased metabolism of cholesterol • *Cholesterol-lowering drugs:* E.g., statins • *Severe or chronic infections:* Increased metabolic demands can decrease cholesterol levels. • *Adrenal or pituitary insufficiency:* Cortisol, adrenal androgens, ACTH, and TSH influence cholesterol metabolism; their deficiency can lead to decreased cholesterol levels. • *Pernicious anemia:* Vitamin B12 is important in lipid metabolism and is deficient. • *Other severe anemias:* Increased erythropoiesis uses up cholesterol to form new RBC membranes. • *Sepsis:* Inflammation can increase the utilization of cholesterol in the immune response and for tissue repair. Inflammatory cytokines can inhibit cholesterol synthesis. • *Burns:* Inflammation • *Hypobetalipoproteinemia:* A genetic deficiency in apolipoprotein B, resulting in decreased LDL production

Low-Density ("BAD") Lipoprotein (LDL)

LDL is the main carrier of cholesterol in the blood, delivering it to various tissues of the body. Although cholesterol is essential for building cell membranes and producing a number of hormones, high levels of LDL cholesterol can accumulate in arterial walls (atherosclerosis) and contribute to cardiovascular disease.

Figure 4-3 lists the causes of increased or decreased LDL.

FIGURE 4-3 LOW-DENSITY LIPOPROTEIN (LDL)

CAUSES OF INCREASED ("BAD") LDL

Increased intake of fats

- High saturated/trans fat/cholesterol diet
- Familial hypercholesterolemia

Increased synthesis of LDL

- *Obesity, sedentary lifestyle:* Reduced activity of enzymes that remove LDL; increased release of fatty acids from adipose tissue leads to increased LDL production.
- *Type 2 diabetes:* Insulin resistance alters lipid metabolism and raises LDL.
- *Nephrotic syndrome:* The liver produces more lipoproteins in response to protein loss in the urine.
- *Cushing's syndrome*: Elevated cortisol stimulates lipoprotein synthesis in the liver.

Decreased clearance of LDL

- *Liver disease:* Ability of the liver to clear LDL is impaired.
- *Alcoholism:* Liver dysfunction impairs LDL metabolism.
- *Hypothyroidism:* Slows down metabolism, including lipid metabolism
- *Growth hormone deficiency:* Decreased lipid breakdown can lead to increased LDL.

CAUSES OF DECREASED LDL

Decreased intake of fats

- *Decreased dietary intake of saturated/trans fats and cholesterol*
- *Malabsorption of fat/cholesterol*

Decreased synthesis of LDL

- *Familial hypobetalipoproteinemia:* Characterized by low LDL
- *Abetalipoproteinemia:* Affects the absorption and transport of lipids, leading to low LDL
- *Cholesterol-lowering drugs*
- *Liver disease*: Can impair cholesterol synthesis and lower LDL

Increased clearance of LDL

- *Exercise*: Can lower LDL and raise HDL
- *Hyperthyroidism*: Stimulates metabolism, including the lowering of LDL
- *Chronic infections and inflammation*: Can lower LDL due to altered metabolism
- *Estrogen replacement therapy*: Can decrease LDL and raise HDL

High-Density ("GOOD") Lipoprotein (HDL)

HDL helps remove excess cholesterol from the tissues, carrying it to the liver in the bloodstream for processing to bile and excretion. Although a high HDL cholesterol level is generally considered beneficial for cardiovascular health, extremely high levels may not be and can contribute to inflammation.

Figure 4-4 lists the causes of an increased or decreased HDL.

FIGURE 4-4 HIGH-DENSITY LIPOPROTEIN (HDL)

CAUSES OF INCREASED HDL

Increased synthesis of HDL

- *Familial HDL lipoproteinemia*
- *Long-term exercise:* Exercise raises HDL
- *Weight loss*
- *Oral estrogens*
- *Alcohol:* Moderate alcohol intake of red wine has been associated with higher HDL.

Other

- *Drugs: Statins*, while associated with lowering LDL, may have some effect in raising HDL. *Estrogen* replacement in postmenopausal women can raise HDL. Omega-3 fatty acids may also raise HDL.
- *Genetic mutation:* Cholesteryl ester transfer protein decreases the level of HDL. A mutation that decreases the activity of this enzyme raises HDL.

CAUSES OF DECREASED HDL

Decreased synthesis of HDL

- *Familial*
- *Chronic liver disease*: HDL is made in the liver and decreases with liver disease.
- *Hypoproteinemia:* E.g., *malnutrition, nephrotic syndrome.* Proteins are needed for HDL synthesis. In nephrotic syndrome, protein is lost in the urine.
- *moking:* Can decrease HDL synthesis and increase HDL breakdown
- *Tangier disease:* A rare inherited mutation in the ABCA1 gene with low HDL and buildup of cholesterol in tissues, with orange-colored tonsils and neuropathy

Triglycerides

Triglycerides are a major source of energy and important in storing excess calories. Triglycerides are carried to cells in VLDL, while cholesterol is carried to cells in LDL. Very Low-Density Lipoprotein (VLDL) is synthesized in the liver, and after delivering triglycerides to fat and muscle cells, VLDL converts to LDL (**Fig. 4-5**).

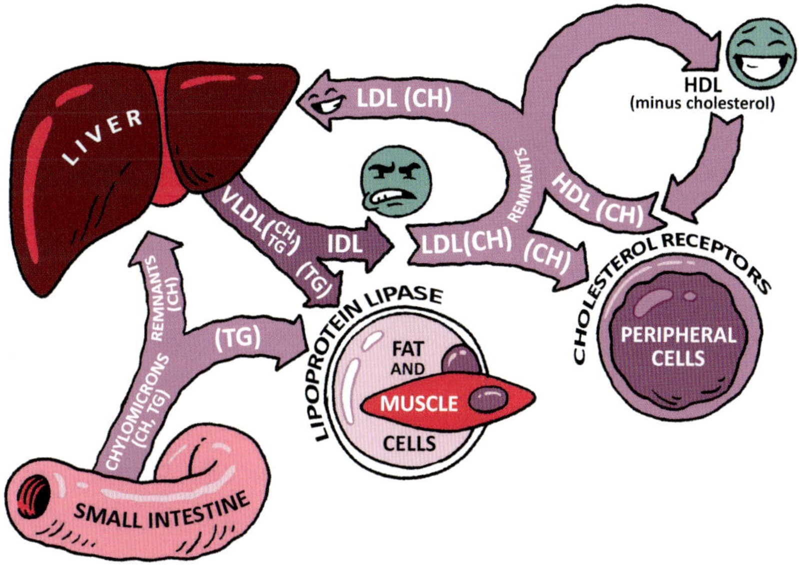

Figure 4-5 *Lipid pathways for triglycerides (TG), LDL, VLDL, and HDL.*

Figures 4-6 and **4-7** list the causes of increased or decreased triglycerides.

FIGURE 4-6 CAUSES OF INCREASED TRIGLYCERIDES
INCREASED DIETARY INTAKE • *Increased ingestion of triglycerides*: Red meat, dairy products • *Increased ingestion of carbohydrates* (which can convert to triglycerides) **INCREASED TRIGLYCERIDE SYNTHESIS** • *Hyperlipidemias,* which can be familial • *Obesity*: Fat accumulates in the liver, where it is converted into triglycerides. • *Glycogen storage disease*: Von Gierke's disease - Increased VLDL synthesis • *Diabetes mellitus*: Insulin lack or resistance leads to release of fatty acids from fatty tissues and their conversion to triglycerides. • *Nephrotic syndrome*: Loss of protein stimulates the liver to produce more proteins, including VLDL, which carries triglycerides. • *Alcohol*: Increases liver production of triglycerides **DECREASED TRIGLYCERIDE LOSS** • *Apoprotein C-II deficiency*: Results in failure to activate lipoprotein lipase, which normally hydrolyzes VLDL triglyceride into free fatty acids and glycerol • *Lipoprotein lipase deficiency:* Lipoprotein lipase breaks down triglycerides.

- *Hypothyroidism:* Decreased catabolism of triglycerides by reducing lipoprotein lipase activity
- *Diabetes mellitus*: Insulin is required to activate lipoprotein lipase, an enzyme that normally breaks down triglycerides.
- *Liver disease*: Can disrupt lipid metabolism
- *Chronic renal failure*: - Insulin resistance leads to decreased ability to activate lipoprotein lipase, needed for triglyceride breakdown.
- *Smoking*: Decreases the activity of lipoprotein lipase, needed for triglyceride breakdown

MEDICATIONS: Can increase liver production of triglycerides or decrease triglyceride breakdown, or both (e.g., steroids; estrogen; diuretics; certain beta-blockers; some antipsychotic medications; retinoids)

FIGURE 4-7 CAUSES OF DECREASED TRIGLYCERIDES

DECREASED DIETARY INTAKE OF TRIGLYCERIDES

- *Malnutrition*
- *Low-fat, low-refined-sugar diet:* Sugar is converted to triglycerides in the liver. Refined sugar has a high glycemic index, where sugar is absorbed more quickly with higher blood sugar levels and ability to convert to triglycerides.
- Increased fiber intake: Decreases the absorption of sugar and fat in the intestines

DECREASED ABSORPTION OF TRIGLYCERIDES

- *Malabsorption of fat*
- *Pancreatic insufficiency*: Reduced pancreatic lipase, which is needed for breakdown and absorption of dietary fats
- *Impaired bile production or secretion*: Bile acids are necessary for the emulsification and absorption of dietary fats.

INCREASED TRIGLYCERIDE LOSS

- *Exercise*: Increases the use of triglycerides for energy
- *Hyperthyroidism*: Increased metabolism of triglycerides

MEDICATIONS: Can increase triglyceride breakdown, inhibit synthesis of triglycerides, or promote the clearance of triglycerides (e.g., fibrates, omega-3 fatty acids, statins, nicotinic acid)

Calculated Lipid Values

Some lipid values are calculated rather than measured, including non-HDL cholesterol, the cholesterol/HDL ratio, and VLDL.

Non-HDL Cholesterol

Calculating non-HDL cholesterol gives a better indication of all atherogenic lipoproteins and thus a better assessment of cardiovascular risk than does LDL alone:

Non-HDL cholesterol = Total cholesterol – HDL cholesterol

Cholesterol/HDL Ratio

A higher cholesterol / HDL ratio indicates a greater risk of heart disease because there is more total cholesterol relative to (good) HDL.

Very Low-Density Lipoprotein (VLDL)

VLDL has a higher proportion of **triglycerides** than LDL and mainly transports *triglycerides* from the liver to the body cells. **LDL** has a higher proportion of **cholesterol** than VLDL and mainly transports *cholesterol* from the liver to the body cells. Knowing VLDL is important in assessing cardiovascular risk, since increased VLDL is associated with an increased risk of cardiovascular disease. VLDL can be measured but is more commonly calculated as:

VLDL cholesterol = Triglycerides / 5

Although LDL cholesterol can be measured directly, it is more commonly calculated by clinical laboratories using the Friedewald formula:

LDL cholesterol = Total cholesterol – HDL cholesterol – (Triglycerides / 5).

However, this formula becomes unreliable when triglyceride levels exceed 400 mg / dL.

Total cholesterol and HDL cholesterol can be measured at any time, including in the non-fasting state. In contrast, triglyceride levels should be assessed after fasting (ideally typically 9-12 hours), as they rise significantly following a fatty meal. As a result, calculated LDL cholesterol may appear falsely low in the non-fasting state—contrary to the common misconception that it might be falsely elevated.

Some advanced lipid panels also measure **Apolipoprotein B** (**ApoB**), the main protein in LDL and VLDL particles. Elevated ApoB is considered a marker of increased atherogenic risk. Another marker, **Lipoprotein(a)** or **Lp(a)**, is a genetic variant of LDL ("bad" cholesterol) with pro-atherogenic, pro-inflammatory, and pro-thrombotic properties.

5

Thyroid Function Tests

The thyroid panel evaluates thyroid function and checks for hyperthyroidism or hypothyroidism.

The panel includes:
Free T3 (triiodothyronine)
Free T4 (thyroxine)
TSH (thyroid-stimulating hormone)

T3 and T4 bind to thyroid binding globulin (TBG), which transports it in the bloodstream. It is the free, rather than the bound T3 and T4 that is metabolically active. High TBG will result in a high total T3 and T4 even though the free forms may be normal. Thus, it is the free forms of T3 and T4 that are more important to measure clinically rather than the total T3 or T4.

Why measure both free T4 and free T3? Is one enough? Free T4 and T3 levels can fluctuate independently. Measuring both gives a more accurate assessment of thyroid function. For instance, Graves' disease or early hyperthyroidism can cause disproportionately higher levels of free T3 than free T4. Autonomously secreting thyroid nodules can also produce more T3 than T4. Thyroid hormone resistance can cause a greater increase in free T4 than T3. It is best to measure both free T4 and free T3.

Free T4 converts to T3, the more active form, within cells, where the T3 binds to thyroid hormone receptors. T4 is a more accurate reflection of thyroid function. About 80-90% thyroid hormones produced are T4; only about 10-20% are T3.

Free Triiodothyronine (T3) and Free Thyroxine (T4)

Figures 5-1 and **5-2** list the causes of increased or decreased T3 and T4.

FIGURE 5-1 CAUSES OF INCREASED FREE T3 AND FREE T4

INTRINSICALLY OVERACTIVE THYROID

Hyperthyroidism

Toxic multinodular goiter

Toxic thyroid adenoma

Acute thyroiditis: Stored T3 and T4 are released during the stage of acute inflammation.

- *Hashimoto's thyroiditis*
- *Subacute granulomatous thyroiditis*
- *Subacute lymphocytic thyroiditis*

EXTRINSIC T3 OR T4

Drug treatment with T3 or T4

Struma ovarii: Ectopic thyroid tissue secretes T3 and T4.

Medication:

- *Amiodarone* can cause hypo- or hyperthyroidism. The excess iodine from am**iod**arone can inflame the thyroid gland and inhibit thyroid hormone synthesis. Alternatively, in some patients the iodine can stimulate the production or release of thyroid hormones.
- *Interferon-alpha* can cause thyroiditis.

INCREASED THYROID-STIMULATING HORMONE (TSH)

Pituitary TSH-secreting tumor (TSH stimulates the thyroid to produce T3 and T4)

Increased thyroid-releasing hormone (TRH) from the hypothalamus (stimulates the pituitary to secrete TSH)

- *Hypothalamic tumor*
- *Hypothalamic inflammation or injury*
- *Congenital hypothalamic dysfunction*
- *Resistance to thyroid hormone* (feedback leads to increased TRH and TSH production to produce more T3 and T4 to compensate)

FIGURE 5-2 CAUSES OF DECREASED FREE T3 AND FREE T4
INTRINSICALLY UNDERACTIVE THYROID ***Hypothyroidism*** ***Surgical removal of thyroid*** ***Burned out thyroiditis*** • *Hashimoto's thyroiditis (later stages)* • *Subacute granulomatous thyroiditis* • *Subacute lymphocytic thyroiditis* • *Reidel's thyroiditis:* Fibrosis replaces normal thyroid tissue. ***Cretinism***
UNDERACTIVE PITUITARY GLAND ***Primary hypopituitarism***: There is a primary defect in pituitary secretion of TSH, so the thyroid is not stimulated.
UNDERACTIVE HYPOTHALAMUS ***Hypothalamic defect***: The hypothalamus does not produce thyroid-releasing hormone (TRH) to stimulate pituitary TSH secretion.
IODINE DEFICIENCY: Iodine is necessary to produce T3.
LIVER DISEASE AND SEVERE ILLNESS*:* The liver normally converts some T4 to T3, so free T3 may be decreased while free T4 may remain normal in liver disease. Thyroid function may still be normal despite the low T3 (euthyroid sick syndrome) since T4 is normal and can still act. The T3 commonly returns to normal with resolution of the illness.

Thyroid-Stimulating Hormone (TSH)

Thyroid-stimulating hormone, produced by the pituitary gland, stimulates the thyroid gland to produce T3 and T4. Measurement of TSH can help tell whether a problem in thyroid hormone secretion is due to a primary thyroid problem or to a secondary problem of pituitary or hypothalamic malfunction. TSH tends to be low in a primary pituitary or hypothalamic underfunction, but high in a primary thyroid underfunction, which stimulates the pituitary by feedback to produce more TSH to compensate (**Figs. 5-3** and **5-4**)

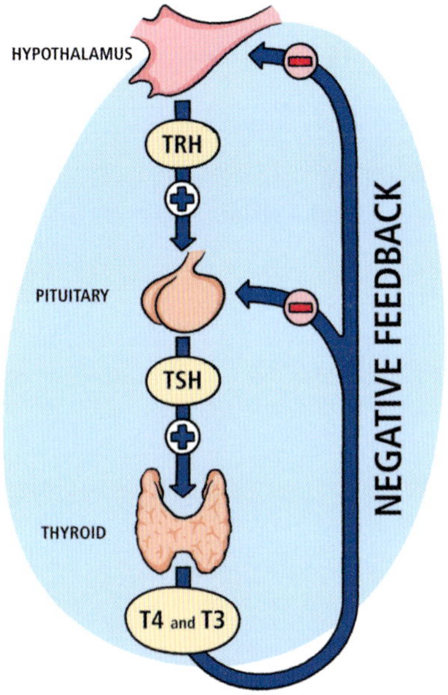

Figure 5-3 *Feedback in the thyroid-hypothalamic-pituitary axis. (Modified from Berkowitz, A. Clinical Pathophysiology Made Ridiculously Simple, Medmaster)*

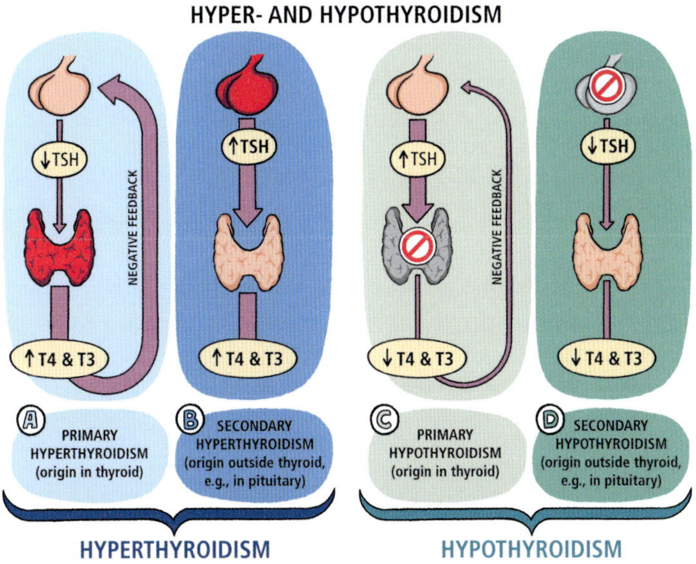

	T4 & T3	TSH
Primary Hyperthyroidism	↑	↓
Secondary Hyperthyroidism	↑	↑
Primary Hypothyroidism	↓	↑
Secondary Hypothyroidism	↓	↓

Figure 5-4 *Types of hyper- and hypothyroidism.*

Figures 5-5 and **5-6** list the causes of increased or decreased thyroid-stimulating hormone (TSH).

FIGURE 5-5 CAUSES OF INCREASED TSH
PRIMARY HYPOTHYROIDISM: The thyroid gland undersecretes T3 and T4, causing feedback that increases TSH secretion by the pituitary gland to try to compensate. ***Burned out thyroiditis*** • *Hashimoto's thyroiditis* • *Subacute granulomatous thyroiditis* • *Subacute lymphocytic thyroiditis* • *Reidel's thyroiditis* ***Congenital hypothyroidism*** ***Agenesis of the thyroid gland*** ***Radioactive iodine administration*** (destroys thyroid tissue) ***Thyroid surgery*** ***Iodine deficiency*** ***Certain medications that impair thyroid function*** (e.g., lithium, amiodarone, interferon-alpha) **SECONDARY HYPERTHYROIDISM:** The pituitary (or an ectopic tumor) oversecretes TSH, causing increased T3 and T4 secretion by the thyroid gland. • *Pituitary TSH-secreting tumor* • *Ectopic TSH-secreting tumor* (e.g., lung, breast) **TERTIARY HYPERTHYROIDISM**: The hypothalamus oversecretes thyrotropin-releasing hormone (TRH), which stimulates the pituitary to secrete TSH. **THYROID HORMONE RESISTANCE**: Causes a compensatory increase in TSH secretion

FIGURE 5-6 CAUSES OF DECREASED TSH
EXCESS T3 OR T4 PRODUCTION OR ADMINISTRATION • *Primary hyperthyroidism*: The thyroid produces too much thyroid hormone, thus suppressing TSH secretion by the pituitary. • *Administration of T3 or T4*: Suppresses TSH secretion by the pituitary • *Certain medications* (e.g., glucocorticoids, dopamine agonists, somatostatin analogs) can inhibit TSH secretion. **PRIMARY DEFECT IN TSH SECRETION** • *Secondary hypothyroidism*: Primary defect in pituitary secretion of TSH • *Tertiary hypothyroidism*: Primary defect in hypothalamus secretion of TRH, so the pituitary does not secrete high enough levels of TSH.

6

Urinalysis

Examination of the urine includes visual, olfactory, chemical, and microscopic analyses (**Figs. 6-1** thru **6-29**).

Visual

Visually, urine may exhibit an unusual color or degree of cloudiness (**Figs. 6-1** and **6-2**).

FIGURE 6-1 CAUSES OF URINE COLORS

STATE OF HYDRATION

- *Clear or light yellow:* Well-hydrated
- *Dark yellow/amber:* Dehydrated

DIET

- *Red/Pink:* Beets, blackberries, rhubarb
- *Orange:* Carrots, vitamin C supplements
- *Blue/Green:* Asparagus, artificial food coloring

DRUGS

Red/Pink: Rifampin, phenazopyridine (Pyridium), senna-containing laxatives
Orange: Phenazopyridine (Pyridium), rifampin, warfarin
Blue/Green: Amitriptyline, indomethacin, propofol, certain dyes used in medical tests
Brown/Tea-colored: Chloroquine, metronidazole, nitrofurantoin, some laxatives.

MEDICAL PROBLEMS

Red/Pink

- *Hematuria:* From urinary tract infection, kidney stones, or trauma
- *Hemoglobinuria:* From hemolysis
- *Myoglobinuria:* From muscle injury

Orange

- *Liver or bile duct problems:* Bilirubinuria
- *Dehydration:* Concentrated urine

Blue/Green

- *Pseudomonas infection*
- *Familial benign hypercalcemia (blue diaper syndrome):* A defect in intestinal absorption of tryptophan, which becomes broken down into indoles in the intestine.

Brown or Tea-colored

- *Liver disease (hepatitis, cirrhosis):* Bilirubin
- *Severe dehydration*
- *Rhabdomyolysis:* Myoglobin from breakdown of muscle

Milky

- *Urinary tract infections*
- *Chyluria:* Lymphatic fluid in the urine

Purple

- *Purple urine bag syndrome:* Associated with bacterial infections in people with urinary catheters. The bacteria convert urinary indoxyl sulfate into indigo and red pigments.

Dark Brown or Cola-colored

- *Acute glomerulonephritis*: From blood and hemoglobin in urine
- *Porphyria*: From porphyrins and their precursors in the urine
- *Certain types of melanoma*: From melanin or melanin precursors in the urine

FIGURE 6-2 CAUSES OF CLOUDY URINE

- *Urinary tract infection:* Bacteria, pus, red blood cells
- *Dehydration*
- *Kidney stones:* Crystals, blood, or WBCs
- *Sexually transmitted infections (STIs):* From bacteria, WBCs, cellular debris, mucus
- *Chyluria:* Lymphatic fluid, sometimes from a fistula between lymphatic system and urinary tract
- *Phosphaturia:* Precipitated excess phosphate in the urine, which may occur in hyperparathyroidism, excess dietary intake of phosphate, renal tubular acidosis (decreased phosphate reabsorption), or certain genetic disorders
- *Proteinuria:* May indicate kidney disease
- *Semen in urine*

Odor

FIGURE 6-3 CAUSES OF URINE ODOR

DIET

Asparagus: Urine may have a strong, sulfur-like odor due to the breakdown of asparagusic acid into sulfur-containing compounds.
Coffee: Often a strong bitter smell to urine
Garlic and onions: Pungent, strong odor due to sulfur compounds.
Fish: Some fish, like tuna, can cause urine to have a fishy smell from compounds like trimethylamine.
Certain spices: Spices like curry can impart a strong, spicy odor to urine.

DEHYDRATION: Urine can have a strong ammonia smell because urea is more concentrated.

MEDICATIONS AND SUPPLEMENTS

- *Vitamins and supplements:* High doses of vitamin B6 (pyridoxine) can cause urine to have a strong, fishy or medicinal odor.
- *Antibiotics:* Some antibiotics, like penicillin, can give a medicinal smell to urine.
- *Chemotherapy:* Can cause urine to have a strong, chemical, medicinal, or metallic odor

MEDICAL CONDITIONS

Urinary tract infections: Bacteria can cause urine to have a foul, pungent or fishy odor.

- *Diabetes:* Sweet-smelling urine due to the presence of glucose and ketones, sometimes described as fruity or like rotten apples
- *Maple syrup urine disease:* Causes urine to have a sweet, maple syrup-like odor from accumulation of branched-chain amino acids

- *Phenylketonuria*: A musty or mousy odor to the urine from the buildup of phenylalanine
- *Liver disease:* Can cause a musty odor to the urine due to the accumulation of metabolic products, like ammonia, dimethyl sulfide, ketones, mercaptans, or phenolic compounds
- *Ketoacidosis:* In diabetic ketoacidosis, ketones cause urine to have a fruity or sweet odor, like acetone or nail polish remover.
- *Bladder fistula:* Can result in urine having a fecal odor due to the connection between the bladder and the intestines.
- *Trimethylaminuria (fish odor syndrome):* Causes a fishy odor in urine due to the body's inability to break down trimethylamine.
- *Pregnancy:* Hormonal changes can sometimes alter the smell of urine, making it stronger or more pungent.

Chemical

FIGURE 6-4 CAUSES OF CHANGE IN URINE PH

DIET

Acidic urine *(low pH)*

- *High protein diets (meat, fish, eggs):* Proteins are metabolized into sulfuric and phosphoric acids, which lower urine pH.
- *Cranberries, plums, and prunes:* Contain organic acids that are excreted in the urine, acidifying it
- *Cheese:* High protein and high calcium and phosphate, which can lower urine pH
- *Citrus fruits (oranges, lemons):* Contain ascorbic acid (vitamin C)

Alkaline urine (high pH)

- *Vegetarian or vegan diets:* Fruits and vegetables are metabolized to bicarbonate, increasing urine pH.
- *Legumes (beans):* High in alkaline substances
- *Dairy products (milk):* Contain calcium and other alkaline substances
- *Baking soda (sodium bicarbonate):* Increases urine pH due to its alkaline nature

MEDICATIONS AND SUPPLEMENTS

Acidic urine* (*low pH)

- *Certain diuretics:* Can lead to a loss of bicarbonate and retention of hydrogen ions, lowering urine pH
- *High doses of vitamin C:* Excess vitamin C is excreted as oxalate and can acidify urine.
- *Ammonium chloride:* Directly acidifies urine

Alkaline urine (high pH)

- *Bicarbonates (antacids):* Increase bicarbonate levels in the blood and urine, raising urine pH

- *Acetazolamide:* Inhibits carbonic anhydrase, leading to bicarbonate excretion and alkaline urine
- *Potassium citrate:* Metabolizes to bicarbonate, raising urine pH

MEDICAL CONDITIONS

***Acidic urine* (low pH)**

- *Metabolic acidosis (diabetic ketoacidosis, lactic acidosis):* Excess acid in the blood is excreted in the urine, lowering pH.
- *Respiratory acidosis:* Retained CO2 increases acid in the blood, lowering urine pH
- *Urinary tract infections:* Certain bacteria (like E. coli, Lactobacillus, and Streptococcus) produce acid by-products such as lactic acid that lower urine pH.
- *Diarrhea:* Loss of bicarbonate
- *Starvation or fasting:* Increased fat metabolism produces ketones, which are acidic.
- *Dehydration:* Concentrates urine, increasing the concentration of acids.

Alkaline urine (high pH)

- *Urinary tract infections:* Certain bacteria (like Proteus, Klebsiella, Pseudomonas, Staph saprophyticus, and Ureaplasma urealyticum) produce urease, which splits urea into ammonia, raising urine pH.
- *Renal tubular acidosis:* Impaired acid secretion or bicarbonate reabsorption results in alkaline urine.
- *Metabolic alkalosis (vomiting, diuretic use)***:** Loss of stomach acid or excessive bicarbonate retention raises urine pH.
- *Respiratory alkalosis (hyperventilation):* Excessive loss of CO2 increases blood pH, causing alkaline urine
- *Overhydration:* Dilutes urine, reducing acid concentration and increasing pH
- *Fanconi syndrome:* Impaired reabsorption of bicarbonate results in its excretion, raising urine pH.

FIGURE 6-5 CAUSES OF INCREASED URINE SPECIFIC GRAVITY

TOO LITTLE WATER EXCRETED

Dehydration

- *Diarrhea*
- *Vomiting*
- *Sweating*
- *Inadequate water intake*

Syndrome of Inappropriate ADH Secretion (SIADH): SIADH results in excess retention of water, with increased urine specific gravity.

Central nervous system causes of SIADH

- *Pituitary tumor* (ADH-secreting)
- *Encephalitis*
- *Meningitis*

- *Trauma to pituitary gland*
- *Brain abscess*
- *Cavernous sinus thrombosis*
- *Acute intermittent porphyria*
- *Multiple sclerosis*
- Subdural hemorrhage
- Subarachnoid hemorrhage
- Hydrocephalus
- Guillain-Barre syndrome

Non-pituitary ADH-secreting tumors
- *Lung cancer*
- *Pancreatic cancer*
- *Thymoma*
- *Lymphoma*
- *Mesothelioma*
- *Bladder cancer*
- *Prostate cancer*

Pulmonary disorders that may cause increased ADH
- *Pneumonia*
- *Lung abscess*
- *Chronic obstructive pulmonary disease:* Dehydration, ADH release
- *Tuberculosis:* Fever, dehydration
- *Aspergillosis:* Fever, dehydration

Various medications may cause water retention

Decreased renal blood flow may cause water retention
- *Congestive heart failure*
- *Renal artery stenosis*
- *Hypotension*

TOO MUCH SOLUTE EXCRETED
- *Diabetes mellitus*
- *Proteinuria*

FIGURE 6-6 CAUSES OF DECREASED URINE SPECIFIC GRAVITY

EXCESS WATER EXCRETION
Excess fluid intake

Diabetes insipidus: Lack of Antidiuretic Hormone (ADH) results in decrease in the kidney's ability to reabsorb water:
- *Central diabetes insipidus*: Pituitary is not excreting ADH.
- *Nephrogenic diabetes insipidus*: Kidney does not respond to ADH.

Renal failure: The kidney does not reabsorb water.

Diuretics

FIGURE 6-7 CAUSES OF INCREASED URINE PROTEIN

EXCESS PROTEIN PRODUCTION

High protein diet

Excess protein production in blood

- *Macroglobulinemia*: Excess globulin production
- *Multiple myeloma*: Bence-Jones protein is excreted.
- *Marked muscle exertion:* Muscle breakdown (rhabdomyolysis) releases myoglobin protein, which appears in the urine.
- *False positive from radiopaque contrast*

Excess protein production in urinary tract

- *Urinary tract infection*
- *Urinary tract hemorrhage*
- *Urinary tract tumor*
- *Prostatic secretions in urine*
- *Vaginal secretions in urine*

DEFECTIVE GLOMERULAR FILTRATION OF PROTEIN: Protein filters through excessively.

- *Nephrotic syndrome*
- *Glomerulonephritis*
- *Diabetic nephropathy*
- *Lupus erythematosus*
- *Pyelonephritis*
- *Malignant hypertension*
- *Preeclampsia*
- *Congestive heart failure*
- *Goodpasture's syndrome*: Antibodies attack lung and kidneys.
- *Polycystic kidneys*
- *Amyloidosis of kidney*
- *Renal vein thrombosis*: Results in renal congestion
- *Orthostatic proteinuria*: Renal congestion may result from changes in anatomical position
- *Nephrotoxic drugs* (heavy metal poisoning; gold salts; arsenicals; acetazolamide; certain antibiotics and other drugs)

FIGURE 6-8 CAUSES OF URINE GLUCOSE (GLYCOSURIA)
HIGH CARBOHYDRATE INTAKE • *Diet*: Increased intake of carbohydrates increases blood sugar. • *Dextrose-containing IVs*
EXCESS PRODUCTION OF GLUCOSE IN THE BODY • *Cushing's syndrome:* Excess cortisol causes increased blood glucose. • *Corticosteroids:* Increase blood glucose by promoting gluconeogenesis and reducing insulin sensitivity, causing hyperglycemia. • *Stress*: Increased adrenaline and cortisol raise blood glucose. • *Pheochromocytoma*: Epinephrine increases blood glucose to provide energy for the "flight or fight" response. • *Glucagonoma:* Glucagon increases glycogenolysis. • *Hyperthyroidism:* An increased metabolic rate raises blood glucose as the body tries to enhance glucose production for energy. • *Pregnancy*: Hormonal changes can affect glucose balance. Increased blood glucose ensures ready availability of glucose for the fetus.
DECREASE REMOVAL OF GLUCOSE FROM THE BLOODSTREAM • *Diabetes mellitus:* Lack of insulin, or insulin resistance. • *Thiazide diuretics:* Impair insulin secretion and sensitivity, raising blood glucose • *Pancreatitis or pancreatic cancer:* Insulin deficiency raises blood glucose. • *Some beta-blockers (e.g., propranolol, metoprolol):* Can reduce insulin secretion from the pancreas and decrease insulin sensitivity. They can also inhibit gluconeogenesis and glycogenolysis. • *Some antipsychotics (e.g., olanzapine, clozapine):* Can lead to weight gain and insulin resistance.
DECREASE RENAL REABSORPTION OF GLUCOSE • *Renal glycosuria:* A defect in the renal tubules' ability to reabsorb glucose into the bloodstream • *Chemical toxicity to kidney:* Mercury, lead, carbon monoxide • *SGLT2 inhibitors (e.g., canagliflozin, dapagliflozin, empagliflozin):* Are used to treat diabetes by reducing glucose reabsorption in the kidneys
ARTIFACTS OF GLUCOSE TESTING • *Contaminants:* E.g., cephalosporins, vitamin C, aspirin, levodopa, cleaning agents, bacteria, and ketones can sometimes cause false positives. • *Other sugars:* Galactose, fructose, or lactose can test positive for glucose on Benedict's test or Clinitest. Test strips that use glucose oxidase/peroxidase are more specific for glucose. • *Vitamin C:* Can give a *false negative* by interfering with glucose oxidase/peroxidase on the test strip. Vitamin C can give a *false positive* with the older Benedict's test or Clinitest. • *Bacterial contamination:* Bacteria eat glucose and can cause a false negative.

FIGURE 6-9 CAUSES OF URINE KETONES

INCREASED PRODUCTION OF KETONES

Increased breakdown of lipids to ketones

- *Diabetic ketoacidosis*
- *Starvation*
- *High fat diet*
- *Low carbohydrate diet*: Fats are broken down for energy.
- *High protein diet*: Fats are broken down for energy.
- *Glycogen storage disease:* With decreased ability to store glycogen or convert it back to glucose, the body compensates by breaking down fats for energy, with the generation of ketones.
- *Conditions with increased body metabolism:* E.g., hyperthyroidism; fever; pregnancy or lactation, because increased metabolic rate breaks down fat with ketone formation

Alcoholism: Ketones are a metabolic product of alcohol.

Isopropyl alcohol ingestion: Produces the ketone acetone as a metabolite

FIGURE 6-10. CAUSES OF URINE BILIRUBIN

INCREASED PRODUCTION OF SOLUBLE (CONJUGATED; DIRECT) BILIRUBIN

Intrahepatic or biliary duct obstruction leads to Increased conjugated bilirubin (excreted in urine).

- *Liver metastases*
- *Drug cholestasis side effect*: Bile cannot be excreted from the hepatocytes.
- *Inflammation of biliary ducts*
- *Gallstones*
- *Tumors*
- *Surgical scarring*
- *Dubin-Johnson Syndrome*: Poor secretion of conjugated bilirubin from hepatic cells
- *Rotor Syndrome*: Poor secretion of conjugated bilirubin from hepatic cells

FIGURE 6-11 CAUSES OF URINE UROBILINOGEN

Urobilinogen is formed in the intestine from bilirubin; some is reabsorbed into the portal circulation where it may be metabolized by the liver or excreted in part by the kidney.

INCREASED BILIRUBIN PRODUCTION:

- *Hemolytic anemia*

FAILURE OF THE LIVER TO REMOVE REABSORBED UROBILINOGEN:

Urobilinogen is then excreted by the kidney.

- *Liver disease*

FIGURE 6-12 CAUSES OF URINE HEME

BLEEDING ORIGINATING IN URINARY TRACT PATHOLOGY

Urinary tract stones

Urinary tract tumor

Urinary tract trauma

Urinary tract infection

Cystitis glandularis: A benign thickening of the bladder wall

Renal disease

- *Glomerulonephritis*: Blood, which normally does not filter through the glomerulus, can now pass through.
- *Acute renal necrosis*
- *Interstitial nephritis*
- *Pyelonephritis*

HEMOLYTIC ANEMIAS: Result in hemoglobinuria

Mechanical hemolysis

- *Artificial heart valve*
- *Disseminated intravascular coagulation*: RBCs are damaged passing over fibrin strands.
- *Thrombotic thrombocytopenic purpura*: RBCs are damaged passing over microthrombi.

Immune hemolysis

Cirrhosis: Leads to splenic congestion and hypersplenism

Sickle cell anemia

Thalassemia

Hemoglobin C disease

Rh-null disease (absence of all Rh antigens on RBCs, red cell fragility and hemolytic anemia)

Hereditary xerocytosis (fragile RBCs, which have reduced deformability)

Congenital dyserythropoietic anemia: Abnormal RBC cell membrane

FACTITIOUS: May produce red-looking urine

- *Myoglobinuria*: Urine dipstix may confuse myoglobin and hemoglobin. Both myoglobin and hemoglobin contain heme.
- *Porphyria:* The porphyrin molecule has structural similarities to heme.

Drugs

- *Phenolphthalein*
- *Deferoxamine*
- *Some phenothiazines*

Contamination of urine specimen with menstrual blood

Nitrites and Urine Leukocyte Esterase

Increased *nitrites* and *leukocyte esterase* are seen on a urine dipstick with urinary tract infections. Increased nitrites suggest a bacterial infection since certain bacteria have an enzyme nitrate reductase, which reduces dietary nitrates to nitrites. The most common bacteria that do this are *E. coli, Klebsiella, Proteus, Enterobacter*, and *Citrobacter*.

Increased *urine leukocyte esterase* can occur in bacterial infections, too, but can also be seen in non-bacterial infections and other inflammatory conditions such as kidney stones and immune-related urinary tract inflammation. Leukocyte esterase is an enzyme produced by white blood cells and can be elevated in conditions that cause a rise in white cells. Contamination with white blood cells from the external genital region can also cause a false positive leukocyte esterase.

Microscopic

Cells

Cells in the urine may be red blood cells, white blood cells, epithelial cells (**Figs. 6-13** thru **6-15**), bacteria, or yeast cells.

FIGURE 6-13 CAUSES OF URINE RBCS (HEMATURIA)

TRAUMA

- *Direct injury to urinary tract*
- *Kidney stones*
- *Exercise-induced hematuria:* Trauma to bladder or kidney; dehydration concentrates urine, so RBCs are detected more easily

URINARY TRACT INFECTION (UTI) OR INFLAMMATION

- *Urethritis*
- *Cystitis, with or without infection*
- *Pyelonephritis*
- *Prostatitis*

OTHER

- *Bladder or kidney cancer*
- *Glomerulonephritis:* Damaged glomeruli allow RBCs to leak through.
- *Benign prostatic hyperplasia (BPH):* Fragile blood vessels in the prostate
- *Menstruation: Sample contamination*
- *Medications (e.g., blood thinners, NSAIDs)*
- *Inherited disorders (e.g., polycystic kidney disease; sickle cell disease)*
- *Urinary tract obstruction*
- *Vascular abnormalities*
- *Blood-clotting disorders*

FIGURE 6-14 CAUSES OF URINE WBCS
GENITOURINARY INFECTIONS OR INFLAMMATION • *Urinary tract infection:* Bacteria, fungi, some viruses, parasites, tuberculosis • *Cystitis:* With or without infection • G*lomerulonephritis* • *Autoimmune disorders:* E.g., lupus nephritis; interstitial nephritis • *Kidney stones*: Inflammation • *Sexually transmitted diseases* • *Prostatitis* • *Bladder cancer* • *Foreign bodies*

FIGURE 6-15 CAUSES OF INCREASED URINE EPITHELIAL CELLS
• *Occasional cells are normal.* • *Contamination from skin or genital region* • *Bladder or urethral infection:* Can shed cells • *Acute renal tubular necrosis:* Renal tubules are damaged by nephrotoxic drugs or hypotension. • *Interstitial nephritis:* Damaged kidney interstitium, often through renotoxic drugs or allergic reactions • *Kidney stones:* Damage to the urinary tract from a passing stone • *Urinary tract tumors:* Can shed cells into the urine • *Chronic kidney disease:* Can shed cells into the urine

Casts

Urinary casts result from the deposition within the renal tubules of debris that outlines the tubules. Casts may be hyaline, granular, RBC, WBC, epithelial, fatty, or waxy. (**Figs. 6-16** thru **6-22**)

FIGURE 6-16 CAUSES OF URINE TUBULAR HYALINE CASTS

Hyaline casts are made of protein and indicate proteinuria. A small number can be normal.

DISEASES WITH PROTEINURIA

- *Dehydration (e.g., from decreased water intake; overexercise; fever):* Concentrates the urine
- *Renal disease:* May cause the kidney to release more protein into the urine
- *Malignant hypertension:* Can damage the kidneys and allow more protein to enter the renal tubules
- *Preeclampsia:* Causes hypertension and proteinuria

- *Congestive heart failure:* Decreased renal perfusion and increased renal venous pressure, which cause increased leakage of proteins into the urine
- *Orthostatic proteinuria*: Renal congestion may result from changes in anatomical position.
- *Nephrotoxic drugs:* Damage the renal tubular cells and increase the excretion of protein, which can form casts

FIGURE 6-17 CAUSES OF URINE TUBULAR GRANULAR CASTS

Granular casts are made of degenerating cellular debris and protein, usually indicating kidney pathology.

DEGENERATION OF WBC OR EPITHELIAL CASTS
- *Acute tubular necrosis*: Damage to renal tubular cells
- *Pyelonephritis:* Produces cellular debris
- *Glomerulonephritis:* Damage to the kidney glomeruli
- *Lupus nephritis*
- *Nephrotoxic drugs*
- *Severe dehydration*
- *Malignant hypertension*
- *Multiple myeloma*: Tubular damage from myeloma proteins
- *Rhabdomyolysis*: Muscle breakdown releases myoglobin, which can cause kidney damage.
- *Marked exercise*: Renal tubule damage from muscle breakdown (rhabdomyolysis) with myoglobin, which can be toxic to renal tubules

FIGURE 6-18 CAUSES OF URINE TUBULAR RED BLOOD CELL CASTS

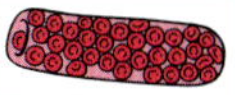

Urine red cell casts indicate glomerular bleeding usually with severe kidney pathology.
- *Glomerulonephritis*
- *Renal trauma*
- *Vasculitis affecting the kidney*
- *Goodpasture's syndrome:* An autoimmune disease with antibodies that attack the lungs and glomeruli
- *Wegener's granulomatosis (granulomatosis with polyangiitis):* An autoimmune disease that attacks blood vessels in the respiratory tract and kidneys, with granulomas.
- *Berger's disease:* IgA antibodies deposit in the glomeruli, causing inflammation, bleeding, and RBC casts.
- *Henoch-Schonlein purpura:* A small vessel vasculitis that can affect the kidney glomeruli

- *Renal vein thrombosis:* Back pressure can injure the kidney, with bleeding into the renal tubules.
- *Acute tubular necrosis*
- *Renal infarct*
- *Sickle cell disease:* When it causes ischemia, infarction, and bleeding in the kidneys
- *Malignant hypertension:* Hypertension can damage the glomeruli
- *Alport syndrome:* A genetic condition where defective type IV collagen in the glomerular basement membrane causes hematuria and RBC casts. (There are also hearing loss and eye abnormalities in the condition.)
- *Bacterial endocarditis:* Can lead to deposition of immune complexes in the glomeruli, causing glomerulonephritis with inflammation, bleeding, and RBC casts

FIGURE 6-19 CAUSES OF URINE TUBULAR WHITE BLOOD CELL CASTS

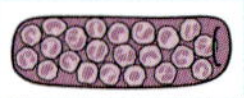

White blood cell casts indicate inflammation or infection within the kidney at or before the renal tubule level.

- *Pyelonephritis*
- *Glomerulonephritis*
- *Interstitial nephritis*
- *Lupus nephritis*
- *Renal tuberculosis*
- *Polycystic kidney disease*
- *Kidney transplant rejection*

FIGURE 6-20 CAUSES OF URINE TUBULAR EPITHELIAL CELL CASTS

Urine epithelial cell casts represent injury to the renal tubules.

- *Glomerulonephritis that extends to the renal tubules*
- *Renal tubule disease*
- *Kidney transplant rejection*
- *Nephrotoxic substances and drugs*
- *Infections that affect the renal tubules*
- *Severe hypertension*
- *Rhabdomyolysis:* Myoglobin from muscle injury can damage renal tubules.

FIGURE 6-21 CAUSES OF URINE TUBULAR FATTY CASTS

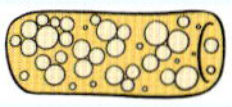

Fatty casts indicate lipiduria and can occur with conditions that cause glomerular damage with leakage of proteins and lipids.

- *Nephrotic syndrome:* From glomerular leakage of lipoproteins
- *Focal segmental glomerulosclerosis:* Glomerular scarring causes protein and lipid leakage.
- *Diabetic nephropathy:* Damages the glomeruli with resulting lipid leakage
- *Membranous nephropathy:* A thickened glomerular membrane from deposition of immune complexes allows leakage of proteins and lipids into the urine.
- *Lupus nephritis*
- *Amyloidosis:* Amyloid deposition in the glomeruli causes leakage of proteins and lipids.
- *Alport syndrome:* A genetic condition where defective type IV collagen in the glomerular basement membrane causes leakage of proteins and lipids. (There are also hearing loss and eye abnormalities in the condition.)

FIGURE 6-22 CAUSES OF URINE TUBULAR WAXY CASTS

Waxy casts indicate chronic kidney disease with tubular atrophy and urine stasis.

- Chronic or end-stage renal disease
- Chronic glomerulonephritis
- Renal amyloidosis
- Diabetic nephropathy
- Hypertensive nephrosclerosis
- Interstitial nephritis

Crystals

Crystals in the urine are caused by the saturation and precipitation of dissolved substances in the urine, commonly from dehydration, metabolic disorders, dietary factors, or underlying renal pathology. Crystals may be amorphous, oxalate, cholesterol, cystine, tyrosine, phosphate, or uric acid (**Figs. 6-23** thru **6-29**)

FIGURE 6-23 CAUSES OF URINE AMORPHOUS CRYSTALS

- *Generally benign*
- *Dehydration*
- *Amorphous urates*: Found in acidic urine, as in high protein diets, which cause acids like sulfuric and phosphoric acid when proteins are metabolized
- *Amorphous phosphates*: Found in alkaline urine and certain bacterial infections. High phosphate diets increase the excretion of phosphates, which are more likely to precipitate in alkaline urine. Bacteria, like Proteus mirabilis, cause an alkaline urine by breaking down urea into ammonia.

FIGURE 6-24 CAUSES OF URINE OXALATE CRYSTALS

- *Oxalate in diet*: Oxalate is found in many foods (e.g., spinach, rhubarb) and oxalate crystals are normal in small amounts.
- *Ethylene glycol ingestion*: Oxalate is a breakdown product of ethylene glycol.
- *Inflammation of lower intestine*: Increased intestinal wall permeability allows more oxalate to be absorbed.
- *Low calcium in lower intestine*: Calcium binds to oxalate and prevents its absorption. Less calcium (e.g., in diet) allows more oxalate to be absorbed.
- *Fatty acid malabsorption*: Unabsorbed fatty acids instead bind to intestinal calcium, leaving less calcium to bind to oxalate, thereby resulting in more oxalate absorption.
- *High intake of vitamin C*: Vitamin C metabolizes to oxalate.
- *Inborn errors in metabolism that result in oxalosis*
- *Parathyroid disease*: Calcium oxalate deposits
- *Renal disease:* Decreased glomerular filtration rate causes increased blood oxalate, which remaining normal renal tubules attempt to excrete, with precipitation of oxalate in the urine. Normal kidneys produce substances (e.g., citrate) that inhibit crystal formation. Renal disease can reduce the formation of these substances, increasing the possibility of crystallization.

FIGURE 6-25 CAUSES OF URINE CHOLESTEROL CRYSTALS

- *Hyperlipidemia*
- *Nephrotic syndrome:* Leakage of lipoproteins, which carry cholesterol
- *Chronic kidney disease:* Leakage of lipoproteins, which carry cholesterol
- *Radiographic media such as iodinated contrast agents*: May resemble cholesterol crystals

FIGURE 6-26 CAUSES OF URINE CYSTINE CRYSTALS

- *Cystinuria: G*enetic mutations in cystine reabsorption

FIGURE 6-27 CAUSES OF URINE TYROSINE CRYSTALS

Tyrosine crystals in the urine point to an underlying metabolic or liver disorder. Tyrosine is a product of protein metabolism.

- *Marked liver disease:* Impaired amino acid metabolism
- *Tyrosinemia:* An inherited disorder where the body cannot break down tyrosine

FIGURE 6-28 CAUSES OF URINE TRIPLE PHOSPHATE CRYSTALS

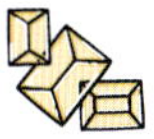

Triple phosphate crystals (struvite crystals, composed of magnesium ammonium phosphate) are typically rectangular prism-shaped. *Calcium phosphate* crystals can appear as flat plates, needles, or rosettes. Both triple phosphate and calcium phosphate crystals are found in alkaline urine, commonly with bacterial infections. Triple phosphate crystals are specifically associated with urease-producing bacteria (e.g., Proteus, Klebsiella, Pseudomonas, Staph saprophyticus). Calcium phosphate crystals often correlate with conditions that promote increased calcium in the urine (e.g., hypercalcemia, kidney stones composed of calcium phosphate).

FIGURE 6-29 CAUSES OF URINE URIC ACID CRYSTALS

Purines metabolize to uric acid. Uric acid crystals are usually rhomboid.

INCREASED DIETARY INTAKE OF PURINES: E.g., organ meats, red meat, certain fish and seafood, beer

PRODUCTION OF TOO MUCH URIC ACID BY THE BODY

- *Breakdown of cells in body*: Released nucleic acids metabolize to uric acid.
- *Metabolic disorders*: E.g., gout
- *Genetic defects in purine metabolism*
- *Dehydration*

Sexually Transmitted Infections

Sexually transmitted infections (STIs) can be detected in several ways:

- Urine tests are used to detect chlamydia and gonorrhea.
- Blood tests can detect HIV, syphilis, hepatitis B, and hepatitis C.
- Swab tests (cervix, urethra, throat, or rectum) are commonly used to detect chlamydia, gonorrhea, and trichomonas.
- A Pap smear is used to screen for papillomavirus (HPV) as well as cervical cancer.

Appendix
Normal Lab Values

Normal laboratory values can differ according to the laboratory and methods used in testing. Evaluation should always be made in the context of the normal values listed by the lab. Also, there may be differences based on age, sex, and ethnicity. The table here lists approximate values.

NORMAL LAB VALUES
Glucose • Fasting: 70-99 mg/dL • 2 hrs postprandial: < 140 mg/dL • Random: < 140 mg/dL for non-diabetics
ELECTROLYTES **Sodium:** 135-145 mmol/L **Potassium**: 3.5-5.0 mmol/L **Chloride**: 98-106 mmol/L **Bicarbonate**: 22-28 mmol/L **CO_2** (total CO_2): 23-29 mmol/L **Calcium**: 8.5-10.2 mg/dL **Magnesium**: 1.7-2.2 mg/dL **Phosphate**: 2.5-4.5 mg/dL Vitamin D: 20-40 ng/mL (lower with darker skin due to decreased sun exposure)
KIDNEY FUNCTION TESTS **BUN**: 7-20 mg/dL Creatinine (higher creatinine with greater muscle mass) • Men: 0.74-1.35 mg/dL • Women: 0.59-1.04 mg/dL
LIVER FUNCTION TESTS **Bilirubin** • Total bilirubin: 0.1-1.2 mg/dL • Direct (conjugated) bilirubin: 0.0-0.3 mg/dL • Indirect (unconjugated) bilirubin: Total bilirubin minus direct bilirubin **ALT (alanine aminotransferase)**: 7-56 U/L

AST (aspartate aminotransferase): 10-40 U/L

ALP (alkaline phosphatase): 44-147 U/L (higher in children and adolescents due to bone growth)

Total protein: 6.0-8.3 g/dL

Albumin: 3.5-5.0 g/dL

Globulin: 2.0-3.5 g/dL

A/G ratio: 1.1-2.2

LIPID PANEL

Total cholesterol (tends to increase with age)

- Desirable: < 200 mg/dL
- Borderline high: 200-239
- High: ≥ 240 mg/dl

LDL (tends to increase with age)

- Optimal: < 100 mg/d:
- Near optimal: 100-129 mg/dL
- Borderline high: 130-159 mg/dL
- High: 160-189 mg/dL
- Very high: ≥ 190 mg/dl

HDL (tends to increase with age)

- Low risk for heart disease: < 40 mg/dL (men); < 50 mg/dL (women)
- High (protective against heart disease): ≥ 60 mg/dL

Triglycerides (tends to increase with age)

- Normal: < 150 mmol/dL
- Borderline: 150-199 mg/dL
- High: 200-499 mg/dL
- Very high: ≥ 500 mg/dL

Non-HDL cholesterol (tends to increase with age)

- Optimal: < 130 mg/dL

Cholesterol/HDL ratio (tends to increase with age)

- Optimal: ≤ 3.5

VLDL: 2-30 mg/dL (tends to increase with age)

THYROID FUNCTION TESTS

Free T3 (triiodothyronine): 2.0-4.4 pg/mL (tends to decrease with age)

Free T4 (thyroxine): 0.8-1.8 ng/dL (tends to decrease with age)

TSH: 0.4-4.0 mIU/L (tends to increase with age)

URINE

Urine pH: 4.5-8.0 (average ~ 6.0)

Urine specific gravity: 1.005 -1.030

RAPID LEARNING AND RETENTION THROUGH THE MEDMASTER SERIES:

BASIC SCIENCES:

CLINICAL NEUROANATOMY MADE RIDICULOUSLY SIMPLE, by S. Goldberg
CLINICAL BIOCHEMISTRY MADE RIDICULOUSLY SIMPLE, by S. Goldberg
ORGANIC CHEMISTRY MADE RIDICULOUSLY SIMPLE, by G.A. Davis
CLINICAL ANATOMY MADE RIDICULOUSLY SIMPLE, by S. Goldberg and H. Ouellette
CLINICAL PHYSIOLOGY MADE RIDICULOUSLY SIMPLE, by S. Goldberg
CLINICAL MICROBIOLOGY MADE RIDICULOUSLY SIMPLE, by M. T. Gladwin, B. Trattler and C.S. Mahan
CLINICAL PHARMACOLOGY MADE RIDICULOUSLY SIMPLE, by J.M. Olson
ACID-BASE, FLUIDS, AND ELECTROLYTES MADE RIDICULOUSLY SIMPLE, by R. Preston
PATHOLOGY MADE RIDICULOUSLY SIMPLE, by A. Zaher
CLINICAL PATHOPHYSIOLOGY MADE RIDICULOUSLY SIMPLE, by A. Berkowitz
CLINICAL BIOSTATISTICS AND EPIDEMIOLOGY MADE RIDICULOUSLY SIMPLE, by A. Weaver and S. Goldberg
CONSCIOUSNESS MADE RIDICULOUSLY SIMPLE: A SERIOUS RESOLUTION OF THE MIND/BODY PROBLEM, by S. Goldberg
CLINICAL GENETICS MADE RIDICULOUSLY SIMPLE, by S. Goldberg

CLINICAL SCIENCES:

OPHTHALMOLOGY MADE RIDICULOUSLY SIMPLE, by S. Goldberg and W. Trattler
PSYCHIATRY MADE RIDICULOUSLY SIMPLE, by J. Nelson, W. Good and M. Ascher
CLINICAL PSYCHOPHARMACOLOGY MADE RIDICULOUSLY SIMPLE, by J. Preston and J. Johnson
BEHAVIORAL MEDICINE MADE RIDICULOUSLY SIMPLE, by F. Seitz and J. Carr
THE FOUR-MINUTE NEUROLOGIC EXAM, by S. Goldberg
CLINICAL RADIOLOGY MADE RIDICULOUSLY SIMPLE, by H. Ouellette and P. Tetreault
THE PRACTITIONER'S POCKET PAL: ULTRA RAPID MEDICAL REFERENCE, by J. Hancock
CLINICAL CARDIOLOGY MADE RIDICULOUSLY SIMPLE, by M.A. Chizner
CARDIAC DRUGS MADE RIDICULOUSLY SIMPLE, by M.A. Chizner and R.E. Chizner
CARDIAC PHYSICAL EXAM MADE RIDICULOUSLY SIMPLE, by M.A. Chizner
ECG INTERPRETATION MADE RIDICULOUSLY SIMPLE, by M.A. Chizner
ORTHOPEDICS MADE RIDICULOUSLY SIMPLE, by P. Tétreault and H. Ouellette
IMMUNOLOGY MADE RIDICULOUSLY SIMPLE, by M. Mahmoudi
ALLERGY AND ASTHMA MADE RIDICULOUSLY SIMPLE, by M. Mahmoudi
RHEUMATOLOGY MADE RIDICULOUSLY SIMPLE, by A.J. Brown
CRITICAL CARE AND HOSPITALIST MEDICINE MADE RIDICULOUSLY SIMPLE, by M. Donahoe and M.T. Gladwin
CLINICAL HEMATOLOGY MADE RIDICULOUSLY SIMPLE, by S. Goldberg
ARE YOU AFRAID OF SNAKES?: A Doctor's Exploration of Alternative Medicine, by C.S. Mahan
WAR AGAINST THE GERMS: EPIDEMICS, MICROORGANISMS, AND BIOWARFARE by S. Goldberg
ENDOCRINOLOGY MADE RIDICULOUSLY SIMPLE, by S. Goldberg

BOARD REVIEW:

USMLE STEP 1 MADE RIDICULOUSLY SIMPLE, by A. Carl
USMLE STEP 2 MADE RIDICULOUSLY SIMPLE, by A. Carl
USMLE STEP 3 MADE RIDICULOUSLY SIMPLE, by A. Carl
NCLEX-RN MADE RIDICULOUSLY SIMPLE, by A. Carl

OTHER:

MEDICAL SPANISH MADE RIDICULOUSLY SIMPLE, by T. Espinoza-Abrams
MED SCHOOL MADE RIDICULOUSLY SIMPLE, by S. Goldberg
MED'TOONS (260 humorous medical cartoons by the author) by S. Goldberg
THE STUTTERER'S APPRENTICE, by A. Splaver

For further information, see http://www.medmaster.net or Email: info@medmaster.net.